Geophysics and Ocean Waves Studies

Edited by Khalid S. Essa, Marcello Di Risio, Daniele Celli and Davide Pasquali

Published in London, United Kingdom

IntechOpen

Supporting open minds since 2005

Geophysics and Ocean Waves Studies
http://dx.doi.org/10.5772/intechopen.87807
Edited by Khalid S. Essa, Marcello Di Risio, Daniele Celli and Davide Pasquali

Contributors
Eman M. Mansour, Ole Bernt Bernt Lile, Jack Dvorkin, Davide Pasquali, Dmitry Chalikov, Khalid S. Essa, Mahmoud Elhussein, Yalin Fan, Paul Hwang, John Yu, Michele Mossa, Diana De Padova, Alessandro Romano, Ahmed Ragab

First published in London, United Kingdom, 2021 by IntechOpen
IntechOpen is the global imprint of INTECHOPEN LIMITED, registered in England and Wales, registration number: 11086078, 5 Princes Gate Court, London, SW7 2QJ, United Kingdom
Printed in Croatia

British Library Cataloguing-in-Publication Data
A catalogue record for this book is available from the British Library

Additional hard and PDF copies can be obtained from orders@intechopen.com

Geophysics and Ocean Waves Studies
Edited by Khalid S. Essa, Marcello Di Risio, Daniele Celli and Davide Pasquali
p. cm.
Print ISBN 978-1-78985-372-8
Online ISBN 978-1-78985-445-9
eBook (PDF) ISBN 978-1-78985-446-6

For EU product safety concerns:
IN TECH d.o.o., Prolaz Marije Krucifikse Kozulić 3, 51000 Rijeka, Croatia, info@intechopen.com or visit our website at intechopen.com.

Meet the editors

Dr. Khalid S. Essa received his B.Sc. with honors (1997), an M.Sc. (2001), and a Ph.D. (2004) in Geophysics from the Faculty of Science, Cairo University. He joined the staff of Cairo University (1997) and was appointed a Research Professor of potential field methods in the Department of Geophysics (2014). He has undertaken affiliated post-doctoral visits to Strasbourg University, France (2018), Charles University in Prague, Czech (2014), and Western Michigan University, USA (2006) and served as an editor and external reviewer for many top journals. He attended several international geophysical conferences in the USA, Australia, and France. He is a member of SEG, AGU, AAPG, EAGE, and EGS. Also, he is a member of the Petroleum and Mineral Resources Research Council, Academy of Scientific Research and Technology, Egypt (2018). He has been awarded the Award of Prof. Nasry Matari Shokry in Applied Geology, Academy of Scientific Research & Technology (2017) and the Award of Cairo University for Scientific Excellence in Interdisciplinary, Multidisciplinary and Future Sciences (2017).

Marcello Di Risio is an Associated Professor (Academic Discipline: Hydraulic Structures, Maritime Engineering, and Hydrology) of the Department of Civil Engineering, Building Architecture and Environmental of the University of L'Aquila. He is the Head of the Laboratory of Environmental and Maritime Hydraulics (LIam) of the same department. He received his Ph.D. in 2005 at the University of Roma Tre. Then, he held a research grant at the University of Rome Tor Vergata and, from 2008 to 2018, he was a researcher at the University of L'Aquila. He has been the chief scientist and coordinator of several research projects, funded by public and private bodies. With many years of experience in the field of hydraulic and maritime construction, his main research topics are: mathematical and experimental modeling of coastal morphodynamic and hydrodynamic phenomena, landslide-generated waves, maritime and hydraulic works, hydraulic risk analysis, real time identification systems of tidal waves, real-time forecasting systems of wave motion and water levels, and development of devices for energy extraction from waves.

Daniele Celli is currently a Post Doctorate student at the Department of Civil, Construction-Architectural and Environmental Engineering (DICEAA) of the University of L'Aquila. In 2015, he received his Master's Degree (cum laude) in Civil Engineering and in 2019 he received his Ph.D. (cum laude) in Risk and Environmental, Territorial and Building Development at the Technical University of Bari. His research interests are focused on coastal structures, water wave generation and propagation, physical and numerical modeling of wave-structure interaction and wave-soil-structure interaction, wave energy assessment and extraction, coastal hydrodynamics, and morphodynamics.

Davide Pasquali is currently a Research Fellow in the Department of Civil, Construction-Architectural and Environmental Engineering (DICEAA) at the University of L'Aquila. In 2011, he received his Master's Degree (cum laude) in Civil Engineering and in 2015 he received his Ph.D. in Civil Engineering at the University of L'Aquila. His research interests are focused on water wave generation and propagation, coastal hydrodynamic and morphodynamic, physical and numerical modeling of wave-structure interaction, wave energy assessment and extraction, risk analysis, and marine sediments transport.

Contents

Preface

Geophysics is the study of the Earth, oceans, atmosphere, and near-space through applying the physics properties and increasing our knowledge about their properties. Geophysics has been useful in many economic applications such as hydrocarbon, ores, and minerals exploration. Also, it is applied in engineering, environmental, archaeological, and deep-Earth investigations.

This book is divided into nine chapters organized into two sections named "Geophysics" and "Ocean Waves Studies".

The first section, "Geophysics", consists of four chapters by scientists and researchers from different countries. It demonstrates the importance of using different geophysics methods in exploration. The first chapter (Essa and Elhussein, 2020) presents a combined formula for gravity and self-potential methods to interpret the buried geologic structures represented by simple models. The second chapter (Lile, 2020) used EM Turam measurements in mining exploration. The third chapter (Dvorkin, 2020) presents the history and basics of rock physics, the effect of pore fluid on rock properties, and the multiple variables affecting rock properties. The fourth chapter (Ragab and Mansour, 2020) demonstrates the enhanced oil recovery phase during oil reservoir production after the water/gas injection and discusses the chemical enhanced oil recovery that reduces the residual oil saturation.

The second section, "Ocean Waves Studies", describes that open seas, enclosed basins, and coastal areas are of utmost importance for human activities. Hence, since the last world war, scientists and engineers have spent much effort in gaining insight on the main (and actually quite fascinating) physical phenomena that occurs in such a water body: surface waves. Long waves, wind waves, infragravity waves, earthquake-induced tsunamis, landslide-induced impulse waves are only a few examples of the wide range of water oscillations that engineers, with the help of scientists, need to face to guide the sustainable use of natural resources. The new sensitivity to climate change and integrated management are only two examples of new challenges to be faced. Mathematical modeling, either analytical or numerical, and experimental investigations are valuable tools that can be used to gain insight into wave generation, propagation, and interaction with the boundaries of water bodies and that are continuously and rapidly improving thanks to technological advances.

The "Ocean Waves Studies" section of this book presents five chapters covering the modeling of wind and surface gravity waves, the storm surge forecast and hindcast, the modeling of the landslide-generated tsunami, and the hydrodynamics of breaking waves.

The section is intended to provide the reader with a detailed description of the latest developments on physical and numerical descriptions of wind-generated and long waves, including some new features discovered in the last few years. The section is organized with the aim to accompany the reader from offshore to nearshore phenomena including a description of wave dissipation including large-scale phenomena (i.e., storm surge and landslide-induced tsunamis).

The first chapter (Chalikov, 2020) in this section describes a set of numerical experiments with a three-dimensional phase-resolving model, aimed to investigate the correct reproduction of surface waves under wind action. Particular attention has been paid by the author to the investigation of the correct reproduction of the energy input and dissipation.

On the theme of numerical simulation of ocean surface waves, and in particular, on the description of input and dissipation source function, the second chapter (Fan et al., 2020) proposes a review on the use of third-generation wave models used in the prediction of tropical cyclone generated surface waves. They also focused on the latest developments in the growing field of coupled atmosphere-wave-ocean modeling with a particular focus on tropical cyclone predictions.

Moving from short to long wave fields, the third chapter (Pasquali, 2020) proposes a review of the storm surge forecast and hindcast in semi-enclosed basins. The main aim of the work is to present a simplified strategy in storm surge modeling using a mixed approach in which results provided by numerical models have been corrected through statistical techniques (i.e., neural network) or by physical parameters (i.e., the inverse barometric effect) to give fast prediction/hindcast with a low computational cost, thus maintaining acceptable reliability.

In the context of long waves, the fourth chapter (Romano, 2020) illustrates a review of physical and numerical modeling of landslide-generated tsunamis with particular attention paid to the most important results achieved in the last ten years. A particular focus is dedicated to the use of Computational Fluid Dynamics (CFD) techniques and approaches in the landslide-generated tsunami modeling highlighting the more recent improvements in the field.

Shifting the attention from the wave description and propagation to the breaking phenomena, the fifth chapter (De Padova and Mossa, 2020) closes the section of the book proposing a Weakly Compressible Smoothed Particles Hydrodynamics (WCSPH) model. They highlight the ability of meshless methods with a k–ε turbulence closure model in describing different breaker types. Moreover, the performances of the proposed model have been validated against a set of dedicated physical experiments.

Finally, this edited book will encourage readers, scientists, researchers, geologists, and geophysicists to look further into the frontier topics of Geophysics and Ocean Waves and inspire new research ways for further novel development.

Khalid S. Essa, Ph.D.
Professor,
Geophysics Department,
Faculty of Science,
Cairo University,
Egypt

**Prof. Marcello Di Risio, Ph.D., Dr. Davide Pasquali, Ph.D.
and Dr. Daniele Celli, Ph.D.**
Environmental and Maritime Hydraulic Laboratory (LIam),
Civil, Construction-Architectural and Environmental Engineering Department,
University of L'Aquila,
Italy

Section 1

Geophysics

Chapter 1

Combined Gravity or Self-Potential Anomaly Formula for Mineral Exploration

Khalid S. Essa and Mahmoud Elhussein

Abstract

A combined gravity and/or self-potential anomaly formula is utilized to estimate the model parameters of the buried geologic structures represented by simple geometric. The simple geometric shapes (spheres, cylinders, and sheets) are not really found but often applied to reduce the nonuniqueness in interpreting the gravity and self-potential data. Numerous approaches through the combined formula such as least squares, Werner deconvolution, and the particle swarm optimization method are used. The application of these methods was demonstrated by applying a synthetic gravity and self-potential example without and with 10% random noise to compare their efficiency in estimating the model parameters of the buried structures. Besides, they were applied to two field data for mineral exploration. The appraised model parameter values from each method were compared together and with those published in literature.

Keywords: gravity and/or self-potential, model parameters, noise, mineral exploration

1. Introduction

Minerals exploration is vital in many countries to increase the income of their people and their economy relies upon discovering minerals. The minerals or ores mined have different variety according to its important in the economy. Geophysical passive method such as gravity and self-potential play an important role in discovering these minerals or ores [1–5]. The gravity method based on measuring the variations in the Earth's gravitational field resulting from the density differences between the subsurface rocks while the self-potential method depended on the electrical potential that develops on the earth's surface due to flow of the natural electrical current on the subsurface [6, 7]. The interpretation of gravity and self-potential data falls on the main two categories as follows: the first category depends on three-dimensional and two-dimensional data elucidation [8–13], the second category is depending using the simple geometric-shaped model such as spheres, cylinders, and sheets which are playing a vital role in interpreting the subsurface structures to reach the priors information that help in more investigations [14–20]. In addition, methods depend on the global optimization algorithms such as genetic algorithm [21–24], particle swarm [25, 26], simulated annealing [27–32], flower pollination [33], memory-based hybrid dragonfly [34], differential evolution [35, 36].

Here, a combined formula for both gravity and self-potential [37] is applied to construct this chapter. Moreover, this formula is used to calculate the buried model parameters, for example in case of self-potential data, the parameters are the electric dipole moment or the amplitude coefficient (K), the polarization angle (θ), the depth (z), the shape (q), and the origin location (x_o) while in case of gravity data, the parameters are the amplitude coefficient (K), the depth (z), the shape (q), and the origin location (x_o) for the buried simple-geometric shapes. Three approaches are suggested to interpret the gravity or self-potential anomaly profile through the combined formula. These methods are least squares, Werner deconvolution, and the particle swarm optimization. The advantage of each method is demonstrated by applying a synthetic example for gravity and self-potential data without and with a 10% random noise to compare their efficiency in deducing the buried model parameters. In addition, they tested on two field example for mineral exploration.

2. The suggested combined gravity or self-potential formula

Firstly, the gravity anomaly formula due to simple geometric shapes is [15, 16, 18]

$$g(x_i, z, q) = K \frac{z^m}{\left((x_i - x_o)^2 + z^2\right)^q}. \tag{1}$$

Secondly, the self-potential anomaly formula for the same simple geometric models is [14]

$$V(x_i, z, \theta, q) = K \frac{x cos\theta + z sin\theta}{\left((x_i - x_o)^2 + z^2\right)^q}. \tag{2}$$

In Refs. [1, 37], Eqs. (1) and (2) were used to join together to produce a combined gravity or self-potential formula for the simple geometric structures such as a semi-infinite vertical cylinder, a dike, a horizontal cylinder, and a sphere (**Figure 1**) as follows:

$$J(x_i) = K \frac{c x_i (cos\theta)^n + z^p (sin\theta)^m}{\left((x_i - x_o)^2 + z^2\right)^q}, \tag{3}$$

where K is the amplitude coefficient, which depends on the shape of the buried model, z is the depth, θ is the polarization angle, x_i is the horizontal coordinates, x_o is the origin location of the buried structure, q is the shape (i.e., equals 1.5 for a sphere, 1.0 for a horizontal cylinder, and 0.5 for a semi-infinite vertical cylinder), c, n, p, and m are constants, which depend on the shape [37]. Eq. (3) is the combined formula for interpreting gravity or self-potential data. So, three suggested approaches were applied to estimate the unknown model parameters as follows:

2.1 The least-squares approach

Essa [37] developed this approach, which was relied on solving the problem of finding the depth from the measured data by solving a nonlinear form $F(z) = 0$ by minimizing it in a least-squares sense. After that, the estimated depth was used in estimating other parameters (the polarization angle and the dipole moment for

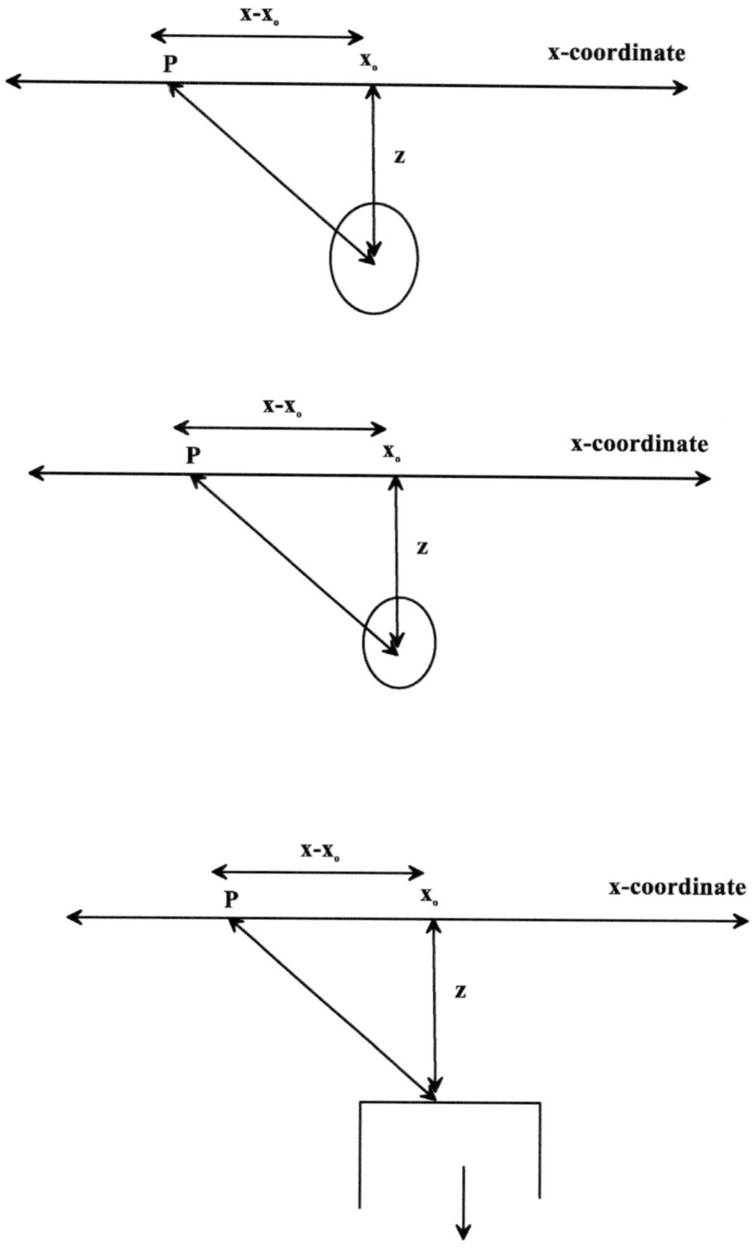

Figure 1.
A sketch diagram for the simple geometric bodies as follows: a sphere model (top panel), a horizontal cylinder model (middle panel), and a semi-infinite vertical cylinder model (bottom panel).

self-potential data or the amplitude coefficient for gravity data) via suggesting the shape of the buried structure (the semi-infinite vertical cylinder, the dike, the horizontal cylinder and the sphere) at the lowest root-mean-squared error. This approach is a semi-automatic because that need assuming the shape of the buried structures (a priori information needed) and applied all observed points in estimating the model parameters.

2.2 Werner deconvolution approach

Werner deconvolution was proposed by Werner in 1953 [38]. This approach is used to estimate mainly the origin location and the depth of the buried structures.

Werner proposed to transform the equation of unknown parameters into a rational function. Eq. (3) can be rewritten in linear form follow:

$$J(x_i)\left((x_i - x_o)^2 + z^2\right)^q - Kcx_i(cos\theta)^n + Kz^p(sin\theta)^m = 0, \quad (4)$$

$$J(x_i)\left((x_i - e_1)^2 + e_2\right)^q - e_3x_i + e_4 = 0. \quad (5)$$

where $e_1 = x_o, e_2 = z^2, e_3 = Kc(cos\theta)^n, e_4 = Kz^p(sin\theta)^m$.

Eq. (5) is linear form in the four variables e_1, e_2, e_3, and e_4, so that a mathematically unique solution can be found for them from evaluating the equation at four points by assuming the shape of the buried structure.

2.3 The particle swarm approach

The particle swarm was suggested by [39] and has many various applications, for example, in geophysics [40–42]. For more detail in this approach, you find it many published literature [43, 44]. The model parameters values of the unknowns are relied upon the objective function, so that every problem can be resolved. In this approach, the particles represent the parameter which we are invert. In the beginning, each particle has a location and velocity. After that each particle changes its location (P_{best}) at every iteration until reach the optimum location (J_{best}). This operation is done by using the following forms:

$$V_i^{k+1} = c_3V_i^k + c_1rand\left(P_{best} - x_i^{k+1}\right) + c_2rand\left(J_{best} - x_i^{k+1}\right), \quad (6)$$

$$x_i^{k+1} = x_i^k + V_i^{k+1}, \quad (7)$$

where v_i^k is the velocity of the particle i at the kth cycle, x_i^k is the current i modeling at the kth cycle, rand is the random number between [0, 1], c_1 and c_2 are positive constant numbers and equal 2, c_3 is the inertial coefficient which control the velocity of the particle and usually taken less than 1, x_i^k is the positioning of the particle i at the kth cycle.

The five source parameters (K, z, θ, x_o, and q) can be assessed by using the particle swarm approach on the subsequent objective function (Obj):

$$Obj = \sqrt{\frac{\sum_{j=1}^{N}\left(J_j^o - J_j^c\right)^2}{N}}, \quad (8)$$

where N is the data points number, J_j^o is the observed gravity or self-potential anomaly, and J_j^c is the estimated anomaly at the point x_j.

3. Synthetic example

To test the ability of each suggested approach in assessing the buried model parameters for the simple geometric shapes such as spheres, cylinders, and sheets. Two synthetic examples are suggested for these interpretation. First one is belonging to use the gravity data and second is applying the self-potential data.

3.1 Gravity anomaly model

A gravity anomaly of a horizontal cylinder model is generated using the following parameters K = 200 mGal×m, z = 5 m, x_o = 0, q = 1.0, and profile length = 100 m

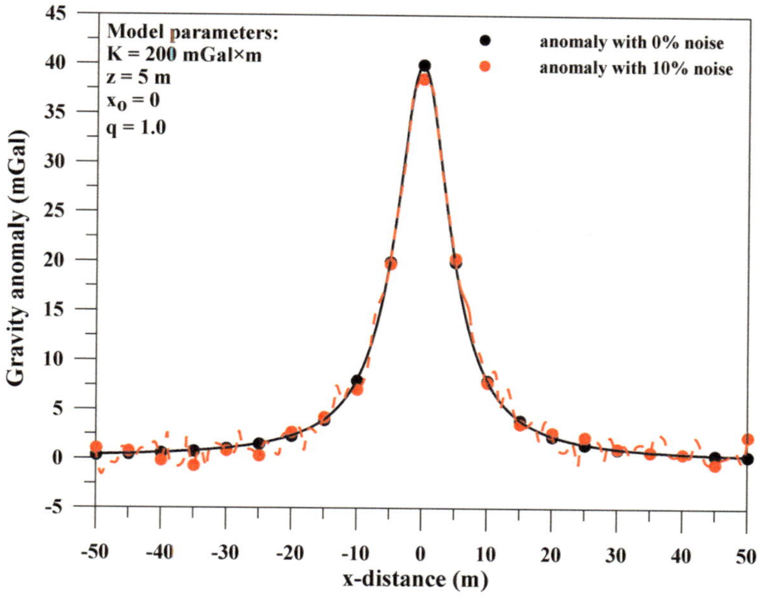

Figure 2.
A gravity model due to horizontal cylinder without and with a 10% of random noise (K = 200 mGal×m, z = 5 m, x_o = 0, q = 1, and profile length = 100 m).

(**Figure 2**). The procedures of interpreting the forward model are done using three steps as follows:

First step: using the least-squares approach to interpret the gravity anomaly yielding from the above mentioned parameters for different s-values for the three suggested shape bodies, i.e., q = 0.5, q = 1.0, and q = 1.5, after that the RMS is

s-values (m)	J(x_j) (mGal)	q = 0.5 K (mGal×m)	q = 0.5 z (m)	q = 1.0 K (mGal×m)	q = 1.0 z (m)	q = 1.5 K (mGal×m^2)	q = 1.5 z (m)
For synthetic data with 0% noise							
+1	38.46	50	2.4	200	5	1100	6.5
-1	38.46						
+2	34.48	50	2.4	200	5	1100	6.5
-2	34.48						
+3	29.41	50	2.4	200	5	1100	6.5
-3	29.41						
+4	24.39	50	2.4	200	5	1100	6.5
-4	24.39						
Average		50	2.4	**200**	**5**	1100	6.5
RMS (mGal)		5.7		**0**		4.1	
For synthetic data with 10% noise							
+1	38.01	48.6	2.3	195.6	4.9	1086.6	6.4
-1	37.92						
+2	34.64	48.5	2.3	196.1	4.9	1082.3	6.5
-2	34.15						
+3	29.15	47.9	2.4	195.8	5.0	1085.4	6.4
-3	28.75						
+4	24.68	48.9	2.3	195.9	5.0	1084.1	6.4
-4	24.40						
Average		48.5	2.3	**195.8**	**4.9**	1084.6	6.4
RMS (mGal)		5.8		**1.0**		4.2	

Table 1.
Numerical results using the least-squares approach for a gravity model due to horizontal cylinder without and with a 10% of random noise (K = 200 mGal×m, z = 5 m, x_o = 0, q = 1, and profile length = 100 m).

Type of body	Parameters	Used ranges	Result	RMS (mGal)
Horizontal cylinder model	For synthetic data with 0% noise			
	K (mGal×m)	100-700	200	0
	z (m)	2-12	5	
	q	0-3	1	
	x_o (m)	-20-50	0	
	For synthetic data with 10% noise			
	K (mGal×m)	100-700	197	0.12
	z (m)	2-12	4.9	
	q	0-3	1.01	
	x_o (m)	-20-50	0.01	

Table 2.
Numerical results using the particle swarm approach for a gravity model due to horizontal cylinder without and with a 10% of random noise (K = 200 mGal×m, z = 5 m, x_o = 0, q = 1, and profile length = 100 m).

calculated to execute the best-fit parameters (happens at the lowest RMS) (**Table 1**). Second step: Werner deconvolution approach is utilized to infer the same gravity data. An 11 clustered solutions to determine in the average evaluated depth (4.9 m) (**Figure 2**). Third step: the particle swarm method is applied to obtain the parameters (**Table 2**).

Moreover, a 10% random noise added to the synthetic gravity data mentioned above (**Figure 2**) to test the efficiency of the suggested approaches in interpreting the gravity data. Also, the three approaches are used for this data as mentioned in

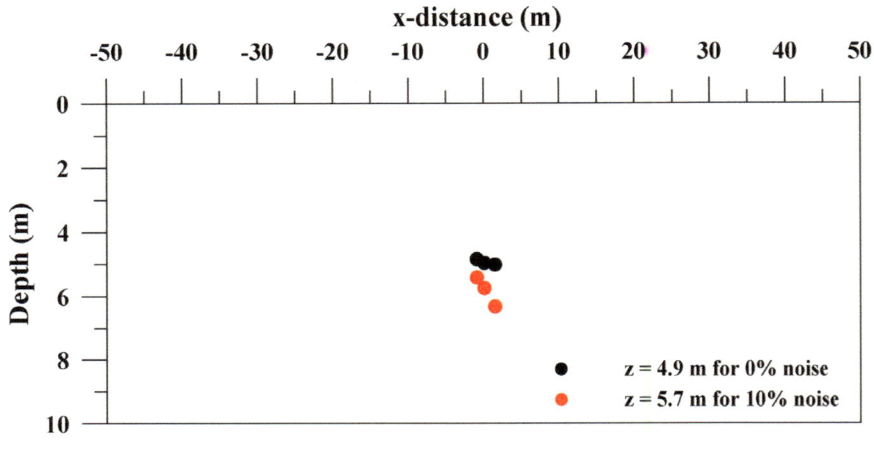

Figure 3.
Werner deconvolution solutions for a gravity model due to horizontal cylinder without and with a 10% of random noise (K = 200 mGal×m, z = 5 m, x_o = 0, q = 1, and profile length = 100 m).

Table 1 (the least-squares approach results), **Figure 3** (Werner deconvolution results), and **Table 2** (the particle swarm results). Finally, the estimated parameters are in all case are in good agreement with the true parameters.

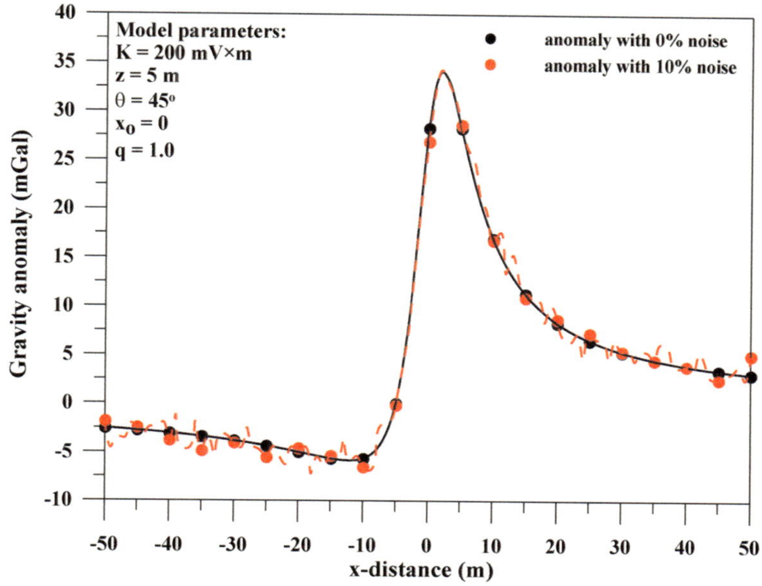

Figure 4.
A self-potential model due to horizontal cylinder without and with a 10% of random noise (K = 200 mV×m, z = 5 m, θ = 45°, x_o = 0, q = 1, and profile length = 100 m).

s-values (m)	$J(x_j)$ (mV)	q = 0.5 K (mV)	q = 0.5 z (m)	q = 0.5 θ (°)	q = 1.0 K (mV×m)	q = 1.0 z (m)	q = 1.0 θ (°)	q = 1.5 K (mV×m²)	q = 1.5 z (m)	q = 1.5 θ (°)
For synthetic data with 0% noise										
+1	38.46	48.9	1.6	67.2	200	5	45	1565.3	6.4	29.7
-1	38.46									
+2	34.48	48.9	1.6	67.2	200	5	45	1565.3	6.4	29.7
-2	34.48									
+3	29.41	48.9	1.6	67.2	200	5	45	1565.3	6.4	29.7
-3	29.41									
+4	24.39	48.9	1.6	67.2	200	5	45	1565.3	6.4	29.7
-4	24.39									
Average		48.9	1.6	67.2	**200**	**5**	**45**	1565.3	6.4	29.7
RMS (mV)		12.8			**0**			4.1		
For synthetic data with 10% noise										
+1	38.01	46.5	1.5	64.8	192.3	4.8	42.1	1512.7	5.9	27.6
-1	37.92									
+2	34.64	47.1	1.6	65.3	195.4	4.8	42.5	1523.9	6.0	27.4
-2	34.15									
+3	29.15	46.7	1.6	66.4	194.3	5.0	43.6	1514.8	6.1	27.1
-3	28.75									
+4	24.68	47.4	1.4	65.9	195.6	4.8	42.8	1531.4	5.8	28.1
-4	24.40									
Average		46.9	1.5	65.6	**195.8**	**4.9**	**4.8**	1520.7	5.9	27.6
RMS (mV)		13.0			**1.1**			4.2		

Table 3.
Numerical results using the least-squares approach for a self-potential model due to horizontal cylinder without and with a 10% of random noise (K = 200 mV×m, z = 5 m, θ = 45°, x_o = 0, q = 1, and profile length = 100 m).

3.2 Self-potential anomaly model

A self-potential anomaly of a horizontal cylinder model is generated using the following parameters K = 200 mV×m, z = 5 m, θ = 45°, q = 1.0, and profile length = 100 m (**Figure 4**). The similar interpretation procedures mentioned above are used

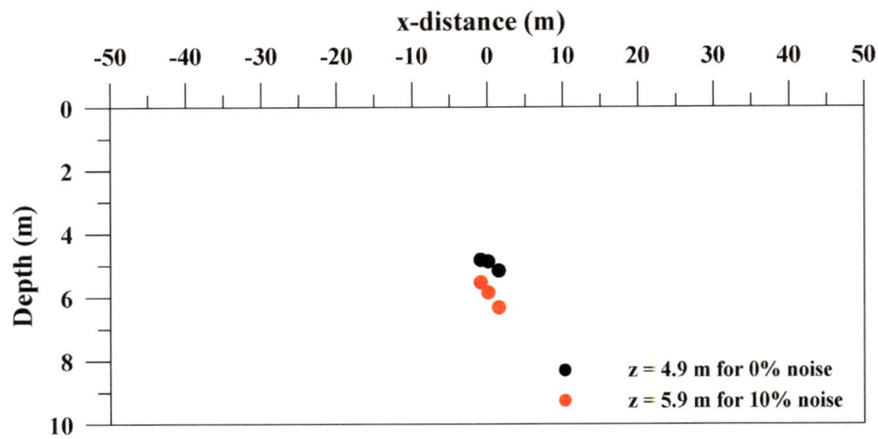

Figure 5.
Werner deconvolution solutions for a self-potential model due to horizontal cylinder without and with a 10% of random noise (K = 200 mV×m, z = 5 m, θ = 45°, x_o = 0, q = 1, and profile length = 100 m).

Type of body	Parameters	Used ranges	Result	RMS (mV)
Horizontal cylinder model	For synthetic data with 0% noise			
	K (mV×m)	100-700	200	0
	z (m)	2-12	5	
	θ (°)	5-85	45	
	q	0-3	1	
	x₀ (m)	-20-50	0	
	For synthetic data with 10% noise			
	K (mV×m)	100-700	195	0.22
	z (m)	2-12	4.9	
	θ (°)	5-85	43.5	
	q	0-3	0.98	
	x₀ (m)	-20-50	-0.02	

Table 4.
Numerical results using the particle swarm approach for a self-potential model due to horizontal cylinder without and with a 10% of random noise (K = 200 mV×m, z = 5 m, θ = 45°, x_o = 0, q = 1, and profile length = 100 m).

as follows: first, the least-squares approach is applied to interpret the self-potential data using various s-values for the three suggested shape bodies, i.e., q = 0.5, q = 1.0, and q = 1.5, after that the RMS is calculated to execute the best-fit parameters (happens at the lowest RMS) (**Table 3**). Secondly, the Werner deconvolution approach is utilized to infer the same self-potential data using 11 clustered solutions to determine in the average evaluated depth (4.9 m) (**Figure 5**). Third step: the particle swarm method is applied to obtain the parameters (**Table 4**). Besides, a 10% random noise was added to this data (**Figure 3**) to test the efficiency of the suggested approaches in interpretation. Furthermore, the results from applying the three approaches are mentioned in **Table 3** (the least-squares approach results), **Figure 5** (Werner deconvolution results), and **Table 4** (the particle swarm results). Finally, the estimated parameters are in all case are in good agreement with the actual parameters.

4. Field examples

The three suggested approaches have been organized to inspect the gravity and self-potential anomalies due to three simple bodies of various structures, e.g., sheets, cylinders, and spheres. Two mineral field examples from India and Turkey have been interpreted to study the reliability of the suggested approaches. The relevant model parameters (K, z, θ, x_o, and q) are evaluated in an integrated way with the existing geological and geophysical results.

4.1 Gravity anomaly of manganese ore body

Figure 6 shows a gravity anomaly was collected over a manganese deposit near Nagpur, India [45]. This gravity profile has a length of 333 m and digitized with an interval of 27 m. This gravity anomaly is subjected to the three interpretation approaches as discussed earlier. Firstly, the interpreted results due to applying the least-squares approach are shown in **Table 5** for various s-values. Besides, the use of

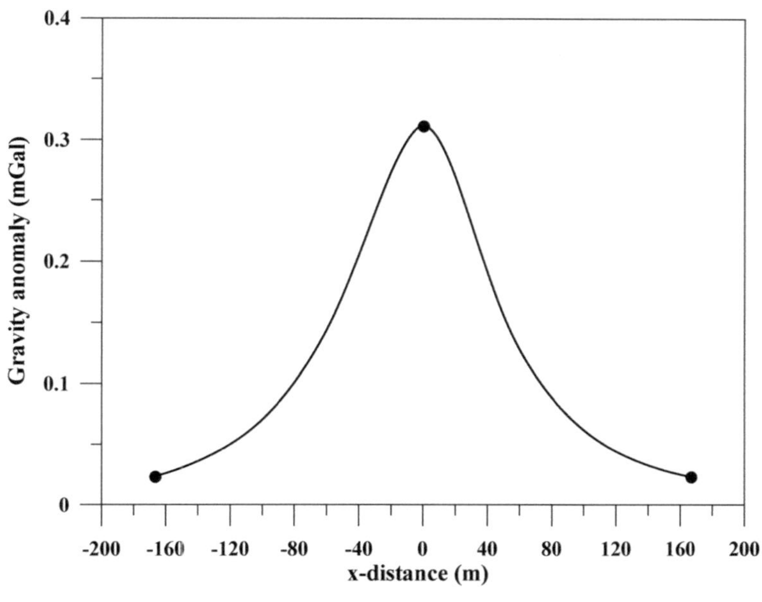

Figure 6.
A gravity anomaly due to a manganese ore body, India.

s-values (m)	$J(x_j)$ (mGal)	q = 0.5		q = 1.0		q = 1.5	
		K (mGal×m)	z (m)	K (mGal×m)	z (m)	K (mGal×m²)	z (m)
+27	0.225	124.6	18.4	24.8	63.2	342.4	112.8
-27	0.233						
+54	0.155	156.1	21.9	25.3	64.3	448.4	123.4
-54	0.139						
+81	0.095	178.3	25.7	22.1	57.4	456.2	124.1
-81	0.107						
Average		153	22	**24.1**	**61.6**	415.7	120.1
RMS (mGal)		40.3		**2.7**		6.2	

Table 5.
Numerical results using the least-squares approach for a manganese field example, India.

Type of body	Parameters	Used ranges	Result	RMS (mGal)
Horizontal cylinder model	K (mGal×m)	1-100	17.3	0.06
	z (m)	1-100	57.8	
	q	0-3	1.0	
	x_o (m)	-10-10	0.8	

Table 6.
Numerical results using the particle swarm approach for a manganese field example, India.

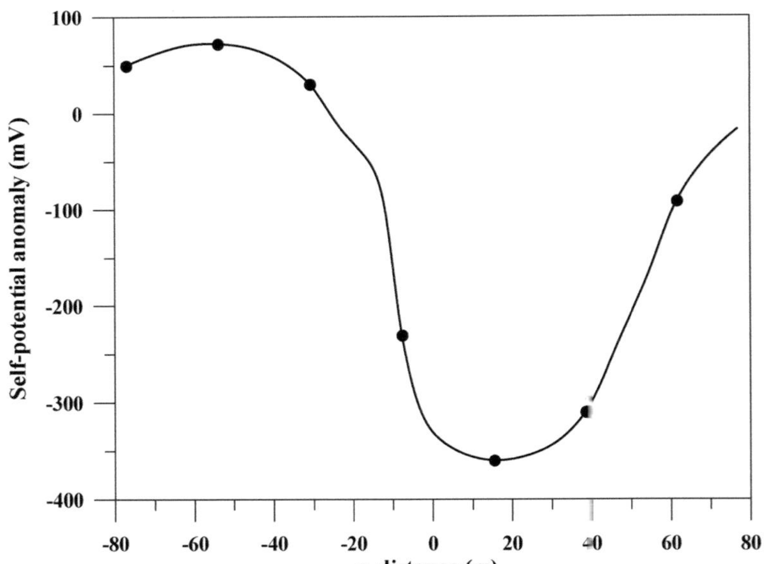

Figure 7.
A self-potential anomaly due to a Weiss copper ore body, Turkey.

s-values (m)	J(xj) (mV)	q = 0.5			q = 1.0			q = 1.5		
		K (mV)	z (m)	θ (°)	K (mV×m)	z (m)	θ (°)	K (mV×m²)	z (m)	θ (°)
+7.7	-352.81	-1148.3	18.4	72.4	-9876.2	63.2	54.6	−472589.6	28.7	38.4
-7.7	-230.57									
+15.4	-359.75	-1192.1	21.9	65.8	-10056.4	64.3	50.9	−471236.1	29.6	37.8
-15.4	-52.78									
+23.1	355.58	-1181.7	25.7	64.9	-9986.3	57.4	49.0	−470893.4	30.1	37.2
-23.1	-15.28									
Average		-1174.0	22.0	67.7	-9972.9	61.6	51.5	**−471236.1**	**29.5**	**37.8**
RMS (mV)		107.9			64.3			**38.0**		

Table 7.
Numerical results using the least-squares approach for a Weiss copper field example, Turkey.

Type of body	Parameters	Used ranges	Result	RMS (mV)
Spherical model	K (mV×m²)	-200000--500000	−471200.3	34.2
	z (m)	10-50	29.7	
	θ (°)	5-85	37.6	
	q	0-3	1.48	
	x_o (m)	-10-10	-2.3	

Table 8.
Numerical results using the particle swarm approach for a Weiss copper field example, Turkey.

Werner deconvolution approach, the interpretive results obtained are z = 56.8 m and x_o = 0.6 m. Finally, the depth and the other model parameters evaluated by using the particle swarm approach are presented in **Table 6**.

4.2 Self-potential anomaly of manganese ore body

Figure 7 demonstrates a self-potential anomaly over a Weiss copper ore body in the Ergani copper district, Turkey [46]. The Weiss self-potential anomaly profile has a length of 144 m and digitized with an interval of 7.7 m. This anomaly has subjected to the three interpretation approaches as discussed earlier. Firstly, the interpreted results due to applying the least-squares approach are shown in **Table 7** for various s-values. Also, the applying of the Werner deconvolution approach, the interpretive results obtained are z = 36.9 m and x_o = −2.1 m. Finally, the depth and the other model parameters evaluated by using the particle swarm approach are presented in **Table 8**.

5. Conclusions

The three geophysical approaches (the least-squares approach, Werner deconvolution approach, and the particle swarm approach) discussed here to interpret gravity or self-potential data using a combined formula for the simple geometric models (spheres, cylinders, and dikes) are stable and give a good results. The stability of these approaches has been confirmed and tested applying two synthetic examples with a 10% and without random noise and two field data for mineral explorations. The estimated parameters in all cases demonstrated the importance of these approaches in interpreting the gravity or self-potential data.

Acknowledgements

The authors would like to thank and express appreciation to Ms. Dolores Kuzelj, Author Service Manager, for her assistance and cooperation in this issue.

Conflict of interest

There is no conflict of interest.

Author details

Khalid S. Essa* and Mahmoud Elhussein
Faculty of Science, Geophysics Department, Cairo University, Giza, Egypt

*Address all correspondence to: khalid_sa_essa@yahoo.com

References

[1] Essa KS. A generalized algorithm for gravity or self-potential data inversion with application to mineral exploration. In: 21th Geophysical conference and Exhibition; Sydney, New South Wales, Australia. 2010

[2] Mehanee S, Essa KS, Smith P. A rapid technique for estimating the depth and width of a two-dimensional plate from self-potential data. Journal of Geophysics and Engineering. 2011;**8**:447-456

[3] Essa KS, Elhussein M. A new approach for the interpretation of self-potential data by 2-D inclined plate. Journal of Applied Geophysics. 2017; **136**:455-461

[4] Essa KS. Minerals. Rijeka, Croatia: InTech d.o.o.; 2019. ISBN: 978-1-83962-682-1

[5] Essa KS, Munschy M. Gravity data interpretation using the particle swarm optimization method with application to mineral exploration. Journal of Earth System Science. 2019;**128**:123

[6] Sato M, Mooney HM. The electrochemical mechanism of sulfide self-potentials. Geophysics. 1960;**25**:226-249

[7] Nettleton LL. Gravity and Magnetics in Oil Prospecting. New York: McGraw-Hill Book Co.; 1976

[8] Zhang J, Zhong B, Zhou X, Dai Y. Gravity anomalies of 2D bodies with variable density contrast. Geophysics. 2001;**66**:809-813

[9] Colangelo G, Lapenna V, Perrone A, Piscitelli S, Telesca L. 2D self-potential tomographies for studying groundwater flows in the Varco d'Izzo landslide (Basilicata, southern Italy). Engineering Geology. 2006;**88**:274-286

[10] Minsley BJ, Sogade J, Morgan FD. Three-dimensional self-potential inversion for subsurface DNAPL contaminant detection at the Savannah River Site, South Carolina. Water Resources Research. 2007;**43**:W04429

[11] Cai H, Xiong B, Zhu Y. 3D Modeling and Inversion of Gravity Data in Exploration Scale, Gravity—Geoscience Applications, Industrial Technology and Quantum Aspect. In: Zouaghi T, editor. Rijeka: IntechOpen; 2017. DOI: 10.5772/intechopen.70961

[12] Biswas A. A review on modeling, inversion and interpretation of self-potential in mineral exploration and tracing paleo-shear zones. Ore Geology Reviews. 2017;**91**:21-56

[13] Biswas A. In: Essa KS, editor. Inversion of Amplitude from the 2-D Analytic Signal of Self-Potential Anomalies, Minerals. Rijeka: IntechOpen; 2019. DOI: 10.5772/intechopen.79111

[14] Essa KS, Mehanee S, Smith P. A new inversion algorithm for estimating the best fitting parameters of some geometrically simple body from measured self-potential anomalies. Exploration Geophysics. 2008;**39**: 155-163

[15] Essa KS. A fast least-squares method for inverse modeling of gravity anomaly profiles due simple geometric-shaped structures. In: Near Surface Geoscience 2012—18th European Meeting of Environmental and Engineering Geophysics; Paris, France. 2012

[16] Essa KS. A fast interpretation method for inverse modelling of residual gravity anomalies caused by simple geometry. Journal of Geological Research. 2012;**2012**: Article ID 327037

[17] Biswas A, Sharma SP. Resolution of multiple sheet-type structures in

self-potential measurement. Journal of Earth System Science. 2014;**123**:809-825

[18] Essa KS. New fast least-squares algorithm for estimating the best-fitting parameters due to simple geometric-structures from gravity anomalies. Journal of Advanced Research. 2014; **5**(1):57-65

[19] Biswas A, Sharma SP. Interpretation of self-potential anomaly over idealized body and analysis of ambiguity using very fast simulated annealing global optimization. Near Surface Geophysics. 2015;**13**:179-195

[20] Mehanee S, Essa KS. 2.5D regularized inversion for the interpretation of residual gravity data by a dipping thin sheet: Numerical examples and case studies with an insight on sensitivity and non-uniqueness. Earth, Planets and Space. 2015;**67**:130

[21] Di Maio R, Rani P, Piegari E, Milano L. Self-potential data inversion through a Genetic-Price algorithm. Computers & Geosciences. 2016;**94**: 86-95

[22] Di Maio R, Piegari E, Rani P. Source depth estimation of self-potential anomalies by spectral methods. Journal of Applied Geophysics. 2017;**136**:315-325

[23] Di Maio R, Piegari E, Rani P, Avella A. Self-potential data inversion through the integration of spectral analysis and tomographic approaches. Geophysical Journal International. 2016; **206**:1204-1220

[24] Di Maio R, Piegari E, Rani P, Carbonari R, Vitagliano E, Milano L. Quantitative interpretation of multiple self-potential anomaly sources by a global optimization approach. Journal of Applied Geophysics. 2019;**162**:152-163

[25] Singh A, Biswas A. Application of global particle swarm optimization for inversion of residual gravity anomalies over geological bodies with idealized geometries. Natural Resources Research. 2016;**25**:297-314

[26] Essa KS, Elhussein M, Youssef MA. Magnetic data interpretation using new techniques: A comparative study. In: Biswas A, Sharma S, editors. Advances in Modeling and Interpretation in Near Surface Geophysics, Springer Geophysics. Cham: Springer; 2020. pp. 263-294

[27] Sharma SP, Biswas A. Interpretation of self-potential anomaly over a 2D inclined structure using very fast simulated-annealing global optimization—An insight about ambiguity. Geophysics. 2013;**78**:WB3-WB15

[28] Biswas A, Sharma SP. Optimization of self-potential interpretation of 2-D inclined sheet-type structures based on very fast simulated annealing and analysis of ambiguity. Journal of Applied Geophysics. 2014;**105**:235-247

[29] Biswas A. Interpretation of gravity and magnetic anomaly over thin sheet-type structure using very fast simulated annealing global optimization technique. Modeling Earth Systems and Environment. 2016;**2**:30

[30] Biswas A, Sharma SP. Interpretation of self-potential anomaly over 2-D inclined thick sheet structures and analysis of uncertainty using very fast simulated annealing global optimization. Acta Geodaetica et Geophysica. 2017;**52**:439-455

[31] Biswas A, Parija MP, Kumar S. Global nonlinear optimization for the interpretation of source parameters from total gradient of gravity and magnetic anomalies caused by thin dyke. Annals of Geophysics. 2017;**60**: G0218

[32] Biswas A. Inversion of source parameters from magnetic anomalies

for mineral/ore deposits exploration using global optimization technique and analysis of uncertainty. Natural Resources Research. 2018;**27**:77-107

[33] Sungkono. Robust interpretation of single and multiple self-potential anomalies via flower pollination algorithm. Arabian Journal of Geoscience. 2020;**13**:100

[34] Ramadhani I, Sungkono S. A new approach to model parameter determination of self-potential data using memory-based hybrid dragonfly algorithm. International Journal on Advanced Science, Engineering and Information Technology. 2019;**9**: 1772-1782

[35] Ekinci YL, Balkaya Ç, Göktürkler G. Parameter estimations from gravity and magnetic anomalies due to deep-seated faults: Differential evolution versus particle swarm optimization. Turkish Journal of Earth Sciences. 2019;**28**: 860-881

[36] Ekinci YL, Balkaya Ç, Göktürkler G. Global optimization of near-surface potential field anomalies through metaheuristics, advances in modeling and interpretation in near surface geophysics. In: Biswas A, Sharma S, editors. Springer Geophysics. Cham: Springer; 2020. pp. 155-188

[37] Essa KS. A new algorithm for gravity or self-potential data interpretation. Journal of Geophysics and Engineering. 2011;**8**:434-446

[38] Werner S. Interpretation of magnetic anomalies at sheet like bodies. Sveriges Geologiska Undersok, Series C, Arsbok. 1953;**43**(6):413-449

[39] Kennedy J, Eberhart R. Particle Swarm Optimization: IEEE Intemational Conference on Neural Networks (Perth, Australia). Vol. IV. Piscataway, NJ: IEEE Service Center; 1995. pp. 1942-1948

[40] Essa KS. A particle swarm optimization method for interpreting self potential anomalies. Journal of Geophysics and Engineering. 2019;**16**: 463-477

[41] Essa KS. Self potential data interpretation utilizing the particle swarm method for the finite 2D inclined dike: Mineralized zones delineation. Acta Geodaetica et Geophysica. 2020. DOI: 10.1007/s40328-020-00289-2

[42] Essa KS, Elhussein M. Interpretation of magnetic data through particle swarm optimization: Mineral exploration cases studies. Natural resources Research. 2020;**29**:521-537

[43] Santos FAM. Inversion of self-potential of idealized bodies anomalies using particle swarm optimization. Computers & Geosciences. 2010;**36**: 1185-1190

[44] Essa KS, Elhussein M. PSO (particle swarm optimization) for interpretation of magnetic anomalies caused by simple geometrical structures. Pure and Applied Geophysics. 2018;**175**:3539-3553

[45] Reddi AGB, Murthy BSR, Kesavanani MA. Compendium of Four Decades of Geophysical Activity in Geological Survey of India. GSI Special Publication No. 36. Geological Survey of India; 1995

[46] Yungul S. Interpretation of spontaneous polarization anomalies caused by spherical ore bodies. Geophysics. 1950;**15**:237-246

Chapter 2

Long Wire Electromagnetic Measurements (Turam EM)

Ole Bernt Lile

Abstract

In Scandinavia, EM measurements have traditionally been popular in sulfide ore exploration. The EM methods using a stationary cable loop or a long wire on the ground surface were called Turam. The wire was grounded by electrodes at the ends. The name, Turam meaning two coils, got the name after the measurement system using two coils measuring the quotient and the phase difference of the vertical field. The measurements were performed in the frequency domain, with frequencies around 400 Hz. Using a large cable loop or a long wire grounded at both ends has advantages as energizing transmitter, which should be utilized in deep exploration. The fall-off rate for the primary field is small, and the electric field can be directed in line with strike direction or the direction of the axis of the mineralization. Examples of the interaction between the energizing cable and the conducting half-space are illustrated by computed models. The grounding points can be shifted with repeated measurements for each grounding position. Both man-made and geological noise can be reduced in this way. Field examples are given in the chapter.

Keywords: EM Turam, electromagnetic, cable configurations, gathered current

1. Introduction

In Scandinavia, electromagnetic measurements have been the dominating class of methods in exploration for metal sulfides since the 1920s. Moving source-moving receiver, Slingram, a lightweight instrument system, became very popular. It was applied for reconnaissance mapping of outcropping conductors under a relatively thin overburden and was successful in the discovery of several orebodies in Sweden.

The other EM method that became popular was Turam. This method used a stationary EM source, large wire loops on the ground, or a long single wire, which was grounded by electrodes at each end. In this chapter, we shall look at the principle of coupling between the wire and a conducting half-space and of coupling between a good conductor (orebody) and the conducting half-space.

2. Long wire electromagnetic measurements (Turam EM)

The electromagnetic method using a long wire or a large loop on the ground surface as energizing device was developed in the 1920s. The advantage using either a large square loop or a single grounded wire is lower fall-off rate of the primary field and thereby larger depth penetration. Current gathering in long conductors is also an advantage in deep exploration [1]. Due to the measurement technique using two coils, the method was called Turam, meaning two coils in old Swedish [2, 3]. There are several publications describing both the theory and field results of this method [4–6]. The measurements were carried out in the frequency domain with a fixed frequency. During the last 40–50 years, there has been a revolutionary development of electronic instrumentation making it possible to work on several frequencies simultaneously in the frequency domain or applying short transmitted pulses and record the response in several channels representing windows with increasing time delays during the pause between the pulses. The time it takes to transmit one pulse and record several channels is in the order of milliseconds, and therefore, a large number of pulses can be used at one measurement station, and the responses in all the channels are stacked. The stacking improves the signal/noise ratio considerably.

Physically, there is no difference between frequency domain measurements with several frequencies and time domain measurements in several time delay channels. Short delay channels correspond to high frequencies in the frequency domain and vice versa.

This article will only deal with measurements in the frequency domain.

3. A long wire on the surface of a conductive half space

A single wire grounded at both ends and connected to an alternating current (AC) generator will set up an alternating magnetic field (primary field), which will induce a voltage in the ground. This will in turn produce inductive currents in the

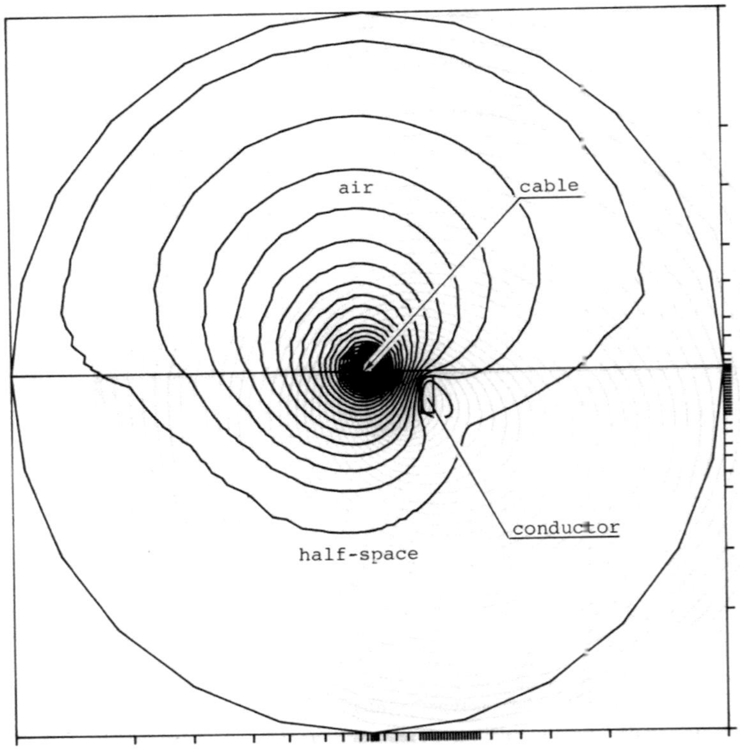

Figure 1.
The total magnetic field from a cable on the surface of a conducting half-space with a conductor [6].

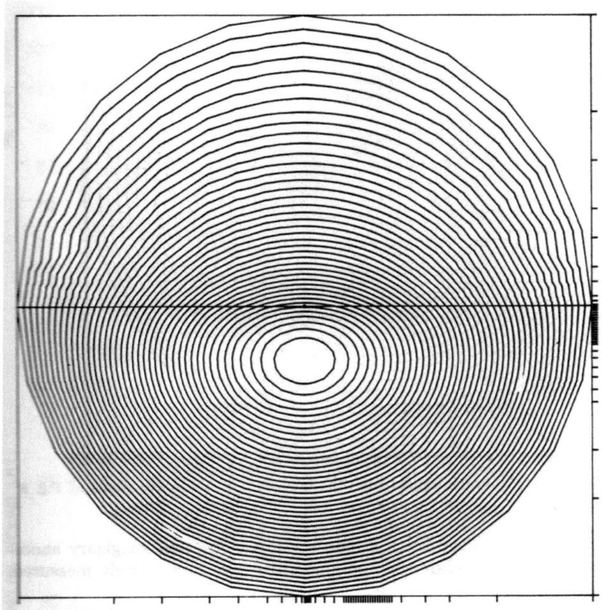

Figure 2.
Secondary magnetic field from induced currents in a conducting half-space [6].

conductive half-space with its own magnetic field (secondary field). This secondary magnetic field is opposing the primary field in the ground, thus weakening the magnetic field from the wire. The resulting field will not penetrate as deep into the ground as in free space and will be phase-shifted relative to the primary field. **Figure 1** shows the resultant field calculated for a two-dimensional model of the half-space and a conductor [6].

The secondary magnetic field alone can be computed and displayed. **Figure 2** shows the induced magnetic field from currents in a conducting half-space without conductor [6]. The induced currents in the ground create a circular pattern of the magnetic field with the vortex below the cable. The depth of the vortex is approximately half the skin depth.

4. Current gathering

Current gathering in long conductors is one of the main advantages with the Turam energizing techniques. This makes deep exploration possible. If it is possible to obtain galvanic contact with a conductor in the ground, one should use this as one of the grounding electrodes of the energizing cable. Lile et al. [6] investigated how gathered currents could be modeled by a two-dimensional computer model. **Figure 3** shows how the anomalous magnetic field lines in the ground are influenced by gathered current in a nearby conductor.

In field measurements, we usually want as much gathered current as possible. Therefore, it is preferable to connect one of the electrodes to the mineralization. However, in some cases, the field from gathered currents can mask anomalies, which may be more interesting. By using grounded electrodes, it is possible to have full control of the current distribution in the half-space and in the mineralization.

In **Figure 4**, the amount of current in the conductive body is 5% of the total current in the half-space. In this case, the edges of the body are outlined. With shallow flat conductors, it is important to have strong induced current at the edges of

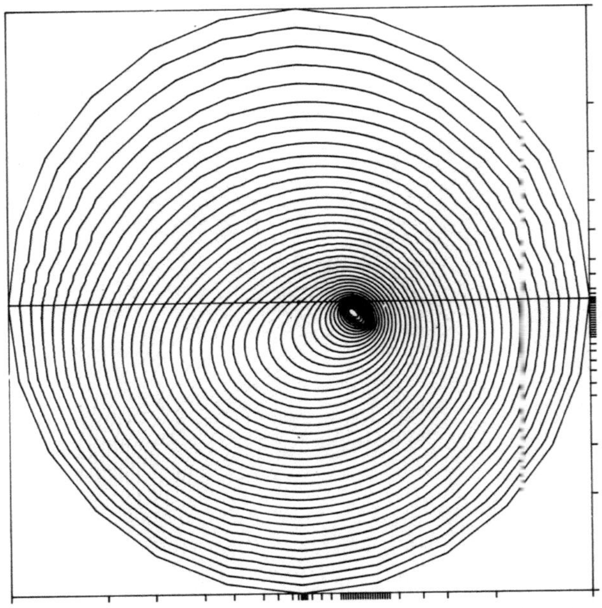

Figure 3.
This figure shows the magnetic field from the induced currents in the half-space and in the long conductor situated in the half-space. The model is the same as shown in ***Figure 1.*** *The good conductor gathers current from the half-space resulting in moving the vortex to the conductor. In this case, the amount of current in the conducting body is 30% of the total amount of current in the half-space [6].*

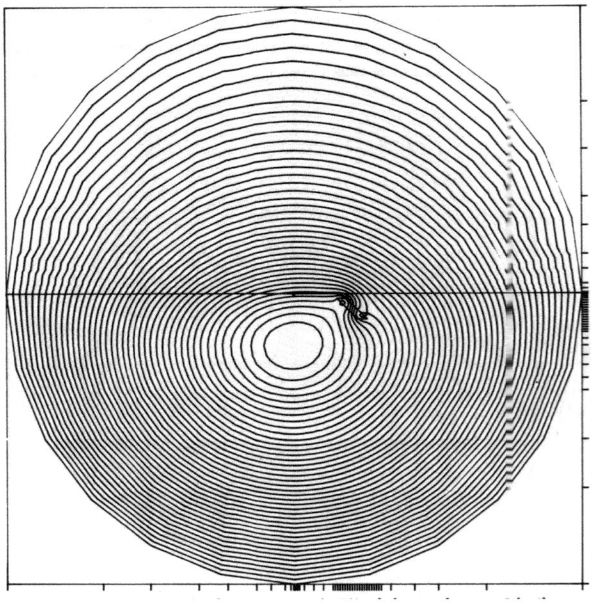

Figure 4.
The amount of gathered current in the conductor is dependent on the conductivity of the body relative to the half-space and the length of the conducting body. In this case, the amount of current in the conducting body is 5% of the total current in the half-space [6].

the body to be able to outline the conductor. The gathered current in the conductor should in such cases be weaker. A cable loop on the ground surface can serve as the transmitter of the primary field.

5. Variable grounding points

In areas where man-made or geological noise is present, interpretation of the data may be difficult. Man-made noise may be conducting structures in the ground or on the surface, which creates anomalies which mask anomalies one is looking for. Geological noise may be varying thickness of conducting overburden or shallow conductive structures in the ground. One method to neutralize such anomalies may be to do measurements at the same stations twice, with variable grounding of the energizing cable.

Lile et al. [1] investigated this technique in a mining area where the orebody was cut by a large fault and the goal was to find the down faulted continuation. The conditions regarding man-made noise were extremely difficult. The exploration area was cut by a main road with heavy traffic, and a railway was situated in parallel and close by the road.

The wire could not be laid parallel to the strike direction because we could not cross the railway and the main road (E6).

In this example, we needed to detect the secondary magnetic field from a sulfide conductor at great depth. Long grounded cables were chosen for energizing the half-space in the exploration area.

Figure 5 shows that we could use a deep ventilation shaft to make two grounding electrodes in weak sulfide zones at approximately 350 m depth. In addition, we made two grounding points at the ground surface, one far away grounding to the north (E) and one to the west, close to the mine (E(0)).

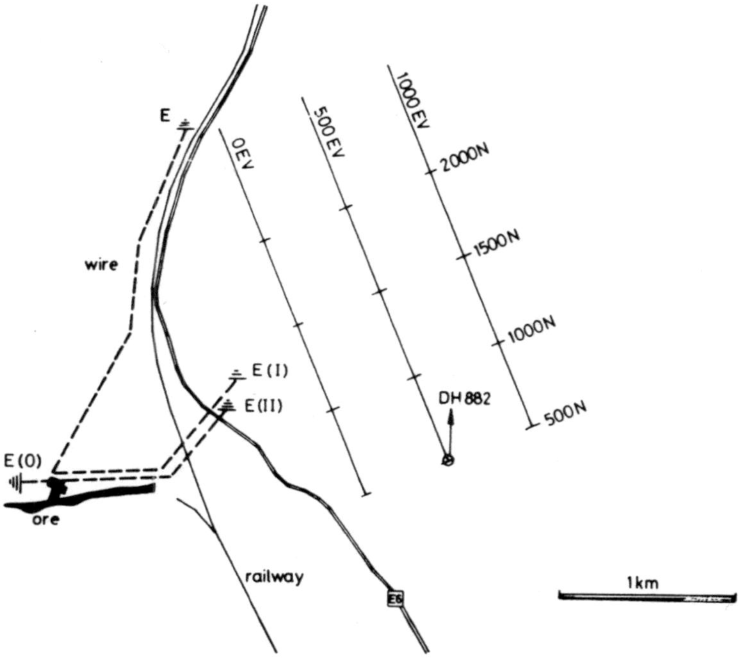

Figure 5.
This is a map showing the situation. The orebody was cut by a NS-fault, dipping 55° to the east, outcropping a couple of 100 m to the west of the railway, and cutting the orebody at 350 m depth. The exploration area was within the grid to the east. The generator was situated at the mine. The picture shows four grounding points. The electrodes to the north (E) and to the west (E(0)) are at the ground surface. The primary magnetic field created by the current in the wire on the surface is needed for the measurements in the grid to the east of the road. The current wire to electrodes E(I) and E(II) goes through a ventilation shaft to a weak sulfide zones at 350 m depth.

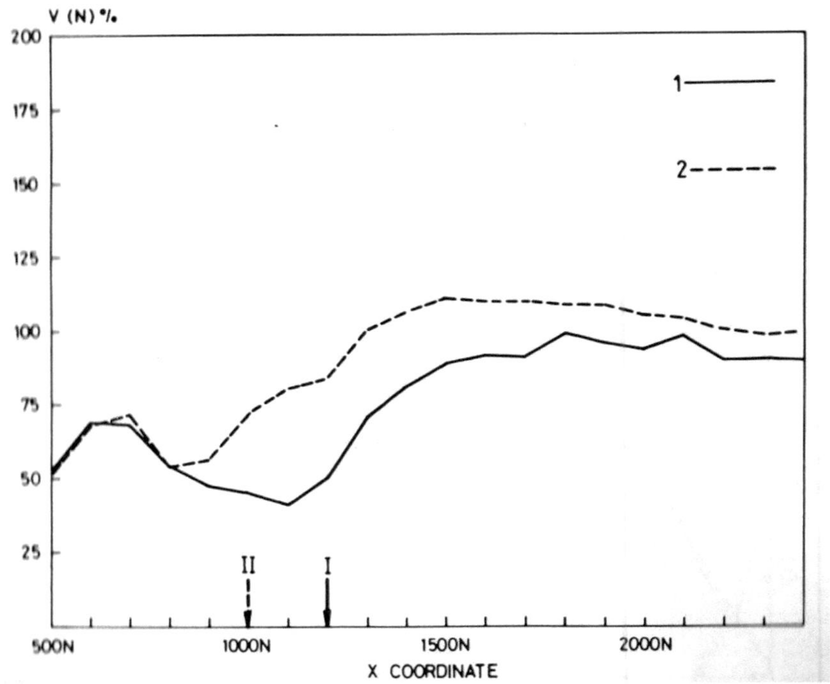

Figure 6.
*Normalized vertical field along profile 600 EV. Curve 1 has been measured with the grounding electrode E(I) and curve 2 with the grounding electrode E(II) at 350 m depth (**Figure 5**). The curves were interpreted as anomalies from conductors at Positions I and II, at depth between 500 and 1000 m. However, a diamond drill hole (DH 882), shown in **Figure 5**, did not hit any conductor, and the exploration was terminated.*

The primary field in the grid came from the long cable between the mine and the electrode to the north (E).

Here only one profile is presented. Profile 600 EV is situated approximately in the middle of the grid. The grounding point to the north (E) was used for all the measurements. A normal vertical field, V(0), was measured with the cable between electrode E and electrode E(0) on the ground surface to the west. At the same points in the grid, the vertical field was measured with the grounding electrodes in the mine, E(I) and E(II), connected to weak sulfide zones. These measurements were then normalized with V(0) as reference and displayed in percentage (**Figure 6**).

Acknowledgements

I am indebted to several partners in these projects I have presented here. O.W. Andersen and R. Pelkonen should especially be mentioned.

Author details

Ole Bernt Lile
Norwegian University of Science and Technology, NTNU, Trondheim, Norway

*Address all correspondence to: obl@ntnu.no

References

[1] Lile OB, Elvebakk H, Berger B. Electromagnetic measurements with variable energizing cable grounding points. Geophysical Transactions. 1984;**30**(2):279-288

[2] Sundberg K. Electrical prospecting in Sweden. Sveriges Geol. Undersökn., Arsbok; 1925

[3] Hedström H. Phase Measurements in Electrical Prospecting. AIMME, Tech. Publ.; 1937. p. 827

[4] Parasnis DS. Mining Geophysics. Amsterdam: Elsevier; 1966. pp. 356

[5] Parasnis DS. Principles of Applied Geophysics. 5th ed. London: Chapman & Hall; 1997. pp. 429

[6] Lile OB, Pelkonen R, Andersen OW, Singsaas P. On the importance of gathered current in turam measurements. Geoexploration. 1982;**19**:277-295

Chapter 3

Rock Physics: Recent History and Advances

Jack Dvorkin

Abstract

This chapter presents the basics of rock physics, the science exploring quantitative relations between various properties (attributes) of the holistic object we call natural rock. This chapter includes several sections, starting with the history and basics; proceeding to the effects of the pore fluid on rock properties; discussing several variables that influence the elastic properties of rocks; presenting selected theories that relate the elastic properties to the porosity, mineralogy, and texture of rocks; and introducing the latest development, digital rock physics. Data examples shown here illustrate qualitative reasoning. Equations are presented as well to mathematically express the conceptual theories discussed. Most importantly, rock physics references are listed to help the reader become willing to delve deeper into the topic and start applying rock physics theories, concepts, and ideas to field data.

Keywords: rock physics, elastic-wave velocities, mineralogy, porosity, permeability, effective medium models

1. Introduction: subject of rock physics, background, and brief history

Rock physics is often called a "velocity-porosity" science. The idea behind this name is to predict the elastic-wave velocities in porous rock from its porosity or implement an inverse operation and interpret the velocity measured in a well or using seismic tomography or reflection techniques for the porosity of rock. It is important to mention that the elastic-wave velocities are related to the elastic moduli of rock as follows:

$$V_p = \sqrt{\frac{K + 4/3G}{\rho_b}}; V_s = \sqrt{\frac{G}{\rho_b}}, \tag{1}$$

where V_p and V_s are the P- and S-wave velocities, respectively; K and G are the bulk and shear moduli, respectively; and ρ_b is the bulk density. The latter quantity is related to the total porosity ϕ as

$$\rho_b = (1 - \phi)\rho_s + \phi\rho_f, \tag{2}$$

where ρ_s is the density of the mineral matrix also called the solid component of the rock, while ρ_f is the density of the pore fluid.

IntechOpen

Important elastic constants used in rock physics are the bulk (K), shear (G), and compressional (M) moduli, as well as the P-wave (I_p) and S-wave (I_s) impedances and Poisson's ratio (ν):

$$M = \rho_b V_p^2; \quad G = \rho_b V_s^2; \quad K = M - 4/3 G;$$
$$I_p = \rho_b V_p; \quad I_s = \rho_b V_s; \quad \nu = \frac{1}{2}\frac{(V_p/V_s)^2 - 2}{(V_p/V_s)^2 - 1}. \tag{3}$$

Most of natural rocks contain more than one mineral. In this situation, ρ_s can be computed as the arithmetic average of the densities of the individual components:

$$\rho_s = \sum_{i=1}^{N} f_i \rho_i, \tag{4}$$

where f_i is the volume fraction of the i-th mineral component in the mineral matrix and ρ_i is its density. These individual densities can be found in handbooks, such as Mavko et al. [1]. They can vary between, e.g., 2.58 g/cc in clay and 4.93 g/cc in pyrite.

The same rule applies to the density of the pore fluid:

$$\rho_f = S_w \rho_w + S_o \rho_o + S_g \rho_g, \tag{5}$$

where S_w, S_o, and S_g are the water, oil, and gas saturations in the pore space, respectively, and ρ_w, ρ_o, and ρ_g are the densities of these pore fluid components. Of course, it is required that

$$\sum_{i=1}^{N} f_i = 1 \tag{6}$$

and

$$S_w + S_o + S_g = 1. \tag{7}$$

Because of the link between the elastic-wave velocities and elastic moduli as given by Eq. (1), it is often instructive to relate these elastic moduli to porosity. Such approach opens an avenue to using the so-called effective medium theories where the elastic moduli are theoretically related to porosity and the geometry of rock, referring to the spatial arrangement of pores and grains, as well as shapes of these pores and grains.

It has been discovered early that the velocity and elastic moduli not only depend on porosity, but also on the properties of the mineral frame. A rule of thumb is that at the same porosity, the softer the mineral frame, the smaller the elastic moduli of rock. For example, at the same porosity, rocks containing soft clays have velocities smaller than rocks dominated by stiffer quartz. Hence, rock physics is not only a "velocity-porosity" science but also a "velocity-porosity-mineralogy" science.

The situation becomes more complex if we consider the effects of the pore fluid on the elastic moduli (and velocities) of a porous composite. It is intuitively clear that the less compressible the pore fluid (water versus gas), the stiffer the entire rock, meaning that its bulk modulus is higher. Now we are talking about "velocity-porosity-mineralogy-fluid."

The science of rock physics also includes understanding and quantification of other rock properties, such as hydraulic permeability and electrical resistivity, and their relation to other attributes, namely, porosity, rock texture, and mineralogy.

Generally, contemporary rock physics treats natural rock as a holistic object whose various properties (attributes) are extracted from experiments simulating processes, such as elastic-wave propagation, fluid and electrical transport, nuclear magnetic resonance (NMR), and breakage. We seek a theoretical understanding of interrelations between such attributes and their mathematical quantification. Such relations are also called rock physics models (RPM) or transforms. Needless to say that such quantification has to be "as simple as possible but not simpler."

Finally, the newest branch of rock physics is digital rock physics (DRP) whose mandate is to "image and compute," image rock at the pore scale and digitally simulate various processes within the digital image. For example, simulations of viscous fluid flow yield permeability, simulations of electrical charge transport yield resistivity, and simulations of deformation under stress yield the elastic moduli.

Let us now review some of historic developments in rock physics.

Arguably, the first rock physics velocity-porosity transform was introduced by Wyllie et al. [2]. It simply states that the total P-wave traveltime through rock with porosity ϕ is the sum of the travel times through the mineral and fluid parts of the rock. This is why it is called the time-average equation. In terms of the P-wave velocities, this formulation is

$$\frac{1}{V_p} = \frac{1-\phi}{V_{\text{ps}}} + \frac{\phi}{V_{\text{pf}}}, \tag{8}$$

where V_p is the P-wave velocity, V_{ps} is the velocity in the mineral phase, and V_{pf} is that in the fluid phase. Examples for 100% quartz and 100% dolomite rock are shown in **Figure 1**. Also shown is an example for rock with mixed 50% quartz and 50% dolomite mineralogy. At the same porosity, V_p is highest in stiffer dolomite, lowest in softer quartz, and falls in between for the mixed mineralogy. The pore fluid was water with V_{pf} = 1500 m/s.

Equation (8) is purely empirical in spite of its physically meaningful form. Indeed, in real rock, the mineral and fluid parts are not arranged in layers to enable a simple summation of the respective traveltimes. Still, this equation gives a reasonably accurate approximation for V_p in "fast" sediments as discussed in Mavko et al. [1]. Also note that it can only work for rock with liquid since in vacuum dry rock, V_{pf} = 0. Yet, as have been shown by seismic experiments on the moon, V_p in such sediment is finite.

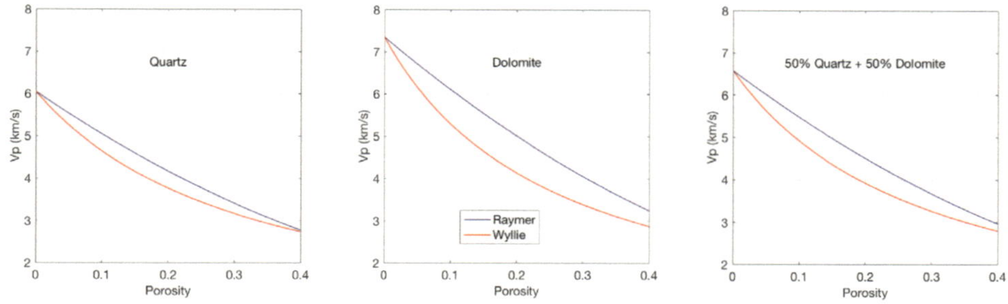

Figure 1.
V_p *versus porosity according to the Wyllie et al. [2] and Raymer et al. [3] transforms for quartz, dolomite, and mixed mineralogy. Legend in the middle refers to all plots.*

Equation (8) has dominated petrophysical interpretation of velocity for porosity for a long time. It gave rise to the so-called sonic porosity computed from wireline velocity data as

$$\phi = \frac{V_p^{-1} - V_{ps}^{-1}}{V_{pf}^{-1} - V_{ps}^{-1}}. \tag{9}$$

The next historic equation was introduced by Raymer et al. [3]:

$$V_p = (1-\phi)^2 V_{ps} + \phi V_{pf}. \tag{10}$$

As Eq. (8), it is purely empirical, derived from wireline data. Still, it is very meaningful as it can be applied to rock with any fluid inside, even where $V_{pf} = 0$. As shown in Dvorkin et al. [4], it is more accurate than the Wyllie et al. [2] time average if applied to "fast" consolidated sediments. Velocity-porosity examples according to this equation are also shown in **Figure 1**.

We conclude this section by presenting equations relating the electrical resistivity to porosity and absolute hydraulic permeability to porosity.

The former transform relates the resistivity R_t of rock fully saturated with conductive fluid (brine) with resistivity R_w as

$$F = \frac{R_t}{R_w} = \frac{1}{\phi^m}, \tag{11}$$

where F is called the formation factor and m is the cementation exponent. In many sandstones m is approximately 2; however it may be much larger in carbonates [1]. **Figure 2** shows experimental data for Fontainebleau sandstone [5] with Eq. (11) curves for m = 1.5, 2.0, and 2.5 superimposed.

At partial brine saturation, $S_w < 1$, the resistivity of rock R_{tS} not only depends on porosity but also on saturation S_w as

$$\frac{R_{tS}}{R_w} = \frac{1}{\phi^m S_w^n} = \frac{F}{S_w^n}, \tag{12}$$

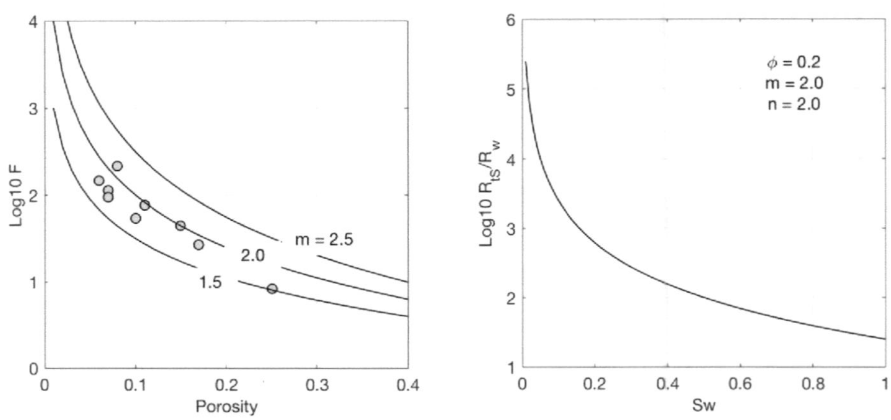

Figure 2.
Left: F *versus porosity according to Eq. (11) for* m = 1.5, 2.0, *and* 2.5 *with Fontainebleau experimental data shown as symbols. Right:* R_{tS}/R_w *ratio versus water saturation for* ϕ = 0.2 *and* m = n = 2.0 *(Eq. (12)).*

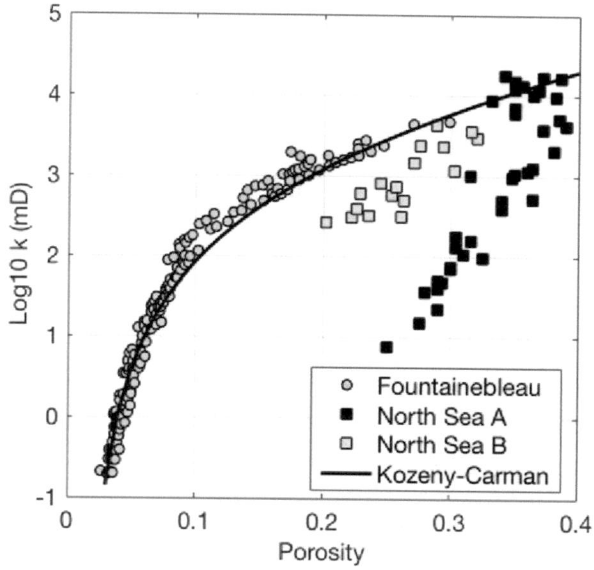

Figure 3.
Permeability versus porosity plots as explained in the text.

where n is the saturation exponent. This exponent is much more elusive than m since laboratory experiments measuring resistivity at partial saturation are scarce. Generally, n should be larger than 1.0 and approach 2.0. An example of R_{tS}/R_w versus S_w is shown in **Figure 2** for porosity 0.2, m = 2.0, and n = 2.0.

Both Eqs. (11) and (12) were discovered by Archie in 1942 [6] and remain the cornerstone of resistivity interpretation for hydrocarbon saturation in the wellbore. Various modifications of these equations dealing with resistivity interpretation in sediments containing clays and shales are discussed in Mavko et al. [1].

The historic absolute permeability prediction equation is called the Kozeny-Carman [7] formula. It is based on an extremely idealized representation of pores as a set of parallel pipes inclined to the direction of pore pressure gradient at an angle α. The tortuosity τ of these pores is defined as

$$\tau = 1/\cos\alpha \geq 1. \tag{13}$$

The permeability k is also a function of the specific surface area S defined as the ratio of surface of the pore space S_{Pore} to the total volume V of the rock sample:

$$S = S_{\text{Pore}}/V. \tag{14}$$

A variable alternative to S is the grain size (or grain diameter) d.

The Kozeny-Carman equation reads [1]

$$k = \frac{1}{2}\frac{\phi^3}{S^2\tau^2} = \frac{1}{72}d^2\frac{\phi^3}{(1-\phi)^2\tau^2}. \tag{15}$$

A modified version of this equation is based on the assumption that k becomes zero not at zero porosity but at a finite and very small porosity value ϕ_p called the percolation porosity:

$$k = \frac{1}{72} d^2 \frac{\left(\phi - \phi_p\right)^3}{\left[1 - \left(\phi - \phi_p\right)\right]^2 \tau^2}. \tag{16}$$

It follows from Eq. (15) that the unit of absolute permeability is length squared. However, traditionally, the permeability unit is Darcy (D) or milli-Darcy (mD). One D is 10^{-13} m^2, while one mD = 10^{-15} m^2.

Figure 3 shows experimental permeability data for Fountainebleau sandstone and two North Sea sand sets with an Eq. (16) curve superimposed for d = 0.25 mm, τ = 2.5, and ϕ_p = 0.02. This theoretical curve matches the Fountainebleau data, while the permeability from the other two datasets falls below this curves. The reason is the varying grain size as discussed in Mavko et al. [1].

2. Effect of pore fluid on elastic properties

Laboratory experiments measuring the elastic-wave velocities in rock often show that the presence of the fluid in the pores strongly affects the elastic properties (**Figure 4**). Such dramatic results, especially for V_p, are in part due to the fact that such experiments are commonly conducted at very high frequencies, on the order of 1 MHz. In this frequency range, the fluid in the pores is "unrelaxed" and acts to strongly reinforce the soft mineral frame, thus increasing the bulk modulus (e.g., [1]).

Arguably, the most important contribution to rock physics is Gassmann's fluid substitution theory [9]. This theory allows us to compute the bulk modulus of porous rock filled with Fluid A if this modulus is known (measured) in the same rock but filled with Fluid B. These derivations were conducted under the assumption that the wave-induced pore pressure oscillations equilibrate within the sample over the wave period, meaning that Gassmann's is a low-frequency theory. Hence, it is applicable at the wireline and seismic frequency ranges. It helps predict the seismic response of rock filled with any hypothetical fluid if it is measured in situ where the pore fluid is known. For example, if the elastic properties of rock are measured in situ in rock 100% filled with water, we can predict these properties in the same rock but filled with oil or gas.

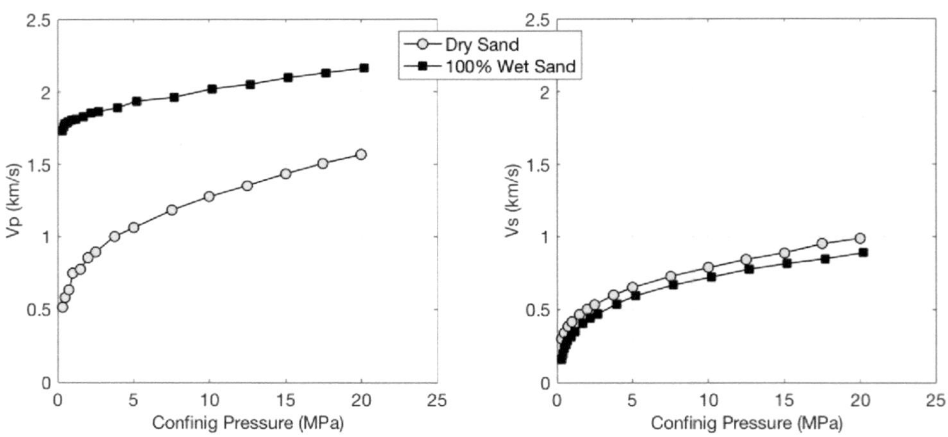

Figure 4.
V$_p$ (left) and V$_s$ (right) of high-porosity unconsolidated sand versus hydrostatic confining pressure. The pore pressure is constant 0.1 MPa. Squares are data obtained in ultrasonic pulse transmission experiments on the water-saturated sample. Circles are for the room-dry sample (after Zimmer [8]).

Gassmann's theory provides the bulk modulus in fluid-saturated rock (K_{Sat}) as a function of the dry rock bulk modulus (K_{Dry}), the bulk modulus of the solid phase (K_s), that of the pore fluid (K_f), and total porosity (ϕ). It assumes that the shear modulus is fluid-independent

$$K_{Sat} = K_s \frac{\phi K_{Dry} - (1+\phi)K_f K_{Dry}/K_s + K_f}{(1-\phi)K_f + \phi K_s - K_f K_{Dry}/K_s}, \quad G_{Sat} = G_{Dry}. \tag{17}$$

The latter equation can be rearranged as follows:

$$K_{Dry} = K_s \frac{1 - (1-\phi)K_{Sat}/K_s - \phi K_{Sat}/K_f}{1 + \phi - \phi K_s/K_f - K_{Sat}/K_s}, \quad G_{Dry} = G_{Sat}. \tag{18}$$

Equations (17) and (18) provide us with a fluid substitution recipe as follows. Assume that we know the bulk modulus K_{SatA} of rock saturated with Fluid A whose bulk modulus is K_{fA} and density is ρ_{fA}. Then from Eq. (17), we obtain:

$$K_{Dry} = K_s \frac{1 - (1-\phi)K_{SatA}/K_s - \phi K_{SatA}/K_{fA}}{1 + \phi - \phi K_s/K_{fA} - K_{SatA}/K_s}. \tag{19}$$

The bulk modulus K_{SatB} of the same rock saturated with Fluid B is (Eq. (17)):

$$K_{\text{SatB}} = K_s \frac{\phi K_{\text{Dry}} - (1+\phi)K_{\text{fB}} K_{\text{Dry}}/K_s + K_{\text{fB}}}{(1-\phi)K_{\text{fB}} + \phi K_s - K_{\text{fB}} K_{\text{Dry}}/K_s}, \tag{20}$$

where K_{fB} is the bulk modulus of Fluid B.

Of course, the shear modulus of the rock remains the same, no matter what fluid it is saturated with.

It is important to remember that the bulk density ρ_b of the rock is also a function of the pore fluid. It depends on the porosity and density of the fluid (ρ_{fA} or ρ_{fB}):

$$\rho_{bB} = \rho_{bA} - \phi\rho_{fA} + \phi\rho_{fB}, \tag{21}$$

where ρ_{bA} and ρ_{bB} are the bulk densities of the rock with the two pore fluids, respectively.

Finally, we can compute the elastic-wave velocities, as well as other seismic attributes, once we know the elastic moduli:

$$V_{pB} = \sqrt{\frac{K_{SatB} + 4/3 G_{Dry}}{\rho_{bB}}}; \quad V_{sB} = \sqrt{\frac{G_{Dry}}{\rho_{bB}}}, \tag{22}$$

and

$$I_{pB} = \rho_{bB} V_{pB}; \quad \nu_B = \frac{1}{2}\frac{(V_{pB}/V_{sB})^2 - 2}{(V_{pB}/V_{sB})^2 - 1}, \tag{23}$$

where I_{pB} and ν_B are the P-wave impedance and Poisson's ratio of the rock filled with Fluid B, respectively. Although the shear modulus G is pore-fluid-independent, V_s is since the bulk density varies with varying fluid.

Let us refer to a later important development in theoretical fluid substitution. It stemmed from the fact that Gassmann's theory [9] requires the knowledge of the bulk modulus that can only be computed using Eq. (1) if both V_p and V_s (and the bulk density ρ_b) are known. In practice, the shear wave velocity may not be available. To address this issue, Mavko et al. [10] derived an approximate (but quite accurate) V_p—only fluid substitution theory that uses the compressional modulus $M = \rho_b V_p^2$ instead of the bulk modulus K. The functional form in this theory is the same as that in Gassmann's:

$$M_{Sat} \approx M_s \frac{\phi M_{Dry} - (1+\phi)K_f M_{Dry}/M_s + K_f}{(1-\phi)K_f + \phi M_s - K_f M_{Dry}/M_s},$$
$$M_{Dry} \approx M_s \frac{1-(1-\phi)M_{Sat}/M_s - \phi M_{Sat}/K_f}{1+\phi - \phi M_s/K_f - M_{Sat}/M_s}. \tag{24}$$

Figure 5 shows an example of the results of fluid substitution (pure water) on the elastic properties of high-porosity sand measured in the laboratory [11] at room-dry conditions. Clearly, the pore fluid has a dramatic effect on Poisson's ratio. Such plots are basis for in situ fluid identification from seismic data.

Let us finally describe the details required in practical fluid substitution, specifically the computation of K_s, ρ_s, K_f, and ρ_f.

The elastic moduli of the multi-mineral rock matrix K_s and G_s can be obtained using Hill's average (e.g., [1]) as

$$K_s = \frac{K_V + K_R}{2}, G_s = \frac{G_V + G_R}{2}, \tag{25}$$

where

$$K_V = \sum_{i=1}^{N} f_i K_i, G_V = \sum_{i=1}^{N} f_i G_i,$$
$$K_R^{-1} = \sum_{i=1}^{N} f_i K_i^{-1}, G_R^{-1} = \sum_{i=1}^{N} f_i G_i^{-1}, \tag{26}$$

where N is the number of the mineral components, f_i is the volume fraction of ith mineral, and K_i and G_i are the bulk and shear moduli of the ith component. The pure-mineral elastic moduli, as well as their densities, can be found in various sources, including Mavko et al. [1].

The bulk modulus of the pore fluid is

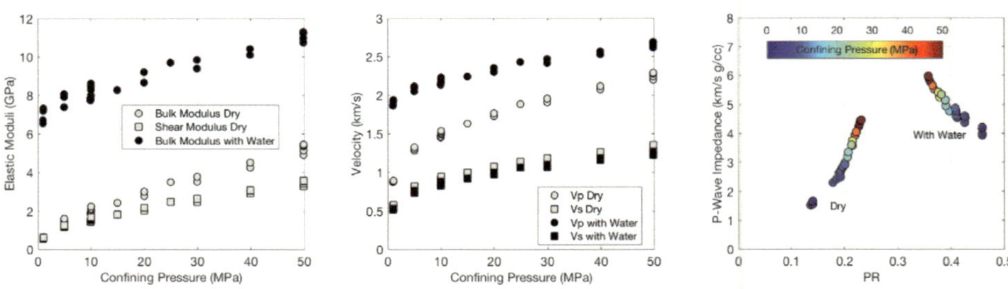

Figure 5.
Sand experimental data and fluid substitution. Left. The bulk and shear moduli versus confining pressure as measured (dry) and water-substituted using Gassmann's theory [9]. Middle. V_p *and* V_s *versus confining pressure as measured (dry) and water-substituted. Right. The P-wave impedance versus Poisson's ratio as measured (dry) and water-substituted, color-coded by the confining pressure.*

$$\frac{1}{K_f} = \frac{S_w}{K_w} + \frac{S_o}{K_o} + \frac{S_g}{K_g}, \tag{27}$$

where K_w, K_o, and K_g are the bulk moduli of water, oil, and gas, respectively. To estimate these moduli, as well as the densities used in Eq. (5), we refer to [12].

3. Variables influencing the elastic properties of rocks

In addition to the pore fluid, there are two more important variables influencing the elastic properties of rocks, their mineralogy and the differential pressure P_{Diff} (or stress) defined as the difference between the confining $P_{\text{Confining}}$ (the overburden) and pore pressure P_{Pore}:

$$P_{\text{Diff}} = P_{\text{Confining}} - P_{\text{Pore}}. \tag{28}$$

Of course there are other influencing factors, such as rock texture (clastics versus carbonates versus unconventional shale), temperature, and diagenetic history. Here we only concentrate on the abovementioned two.

Mineralogy. As an example, let us examine the Han [13] laboratory dataset obtained on a large suite of sandstones with porosity ranging from zero to 30% and clay content between zero and 50%. **Figure 6** shows V_p and V_s versus porosity and color-coded by the clay content.

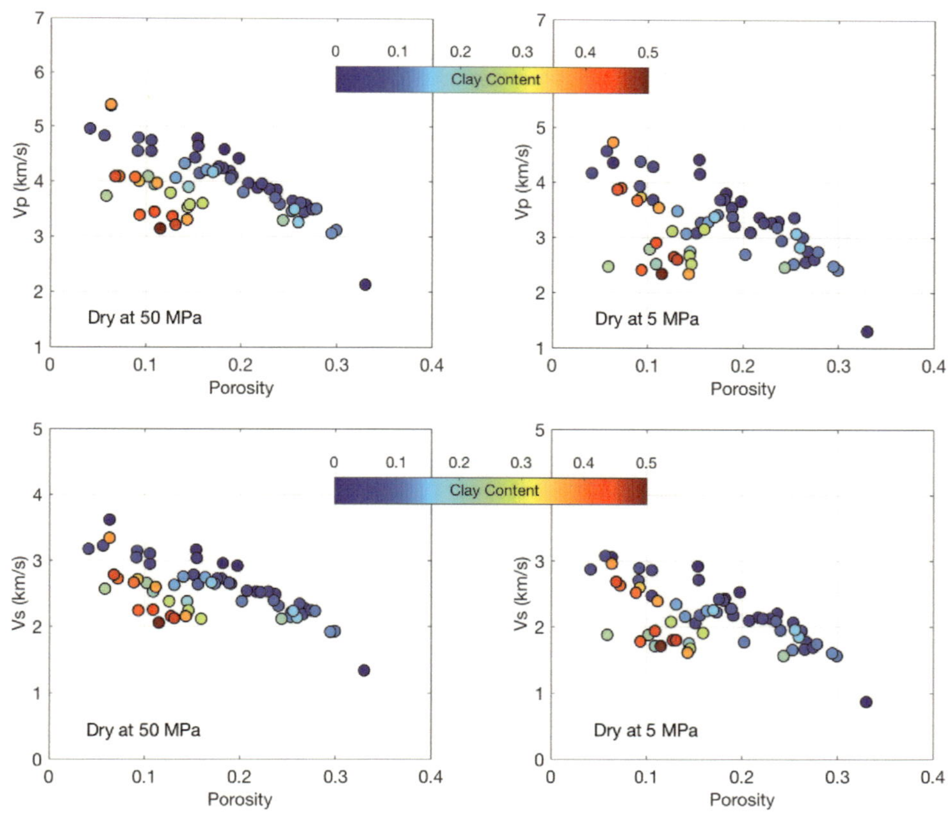

Figure 6.
Dry rock V_p (top) and V_s (bottom) versus porosity, color-coded by the clay content, at confining pressure 50 MPa (left) and 5 MPa (right) (after Han [13]).

Obviously, the clay content plays a dramatic role acting to reduce both V_p and V_s at the same porosity. Also notice that the velocity-porosity-mineralogy trends are much more pronounced at 50 MPa. This is a commonly observed effect due to much clearer manifestations of key rock properties at high confining stress. The high-porosity data point in **Figure 6** at porosity about 0.33 is for unconsolidated Ottawa sand sample. The effect of pressure on its velocities is very strong, similar to what we observe in **Figure 5** for a sand of different provenance.

Another striking example of velocity discrimination due to mineralogy comes from unconventional shale with data obtained by wireline logging in a vertical well (**Figure** 7). The data shown in this figure is for 100% wet rock, obtained by fluid substitution from in situ conditions. The velocity-porosity dataset forms an amorphous cloud (**Figure** 7, top) with both V_p and V_s varying by almost 1.5 km/s at the same porosity. However, as soon as we introduce a third variable, the sum of the clay and kerogen contents, we observe a clear velocity discrimination with the velocity decreasing as the fraction of this softest component of the solid matrix increasing (**Figure** 7, bottom).

The Raymer et al. [3] model also predicts a strong dependence of the velocity on mineralogy (**Figure 8**), as well as the pore fluid, the latter well pronounced at higher porosity.

Stress. The effect of the confining pressure on the velocity in sand can be clearly seen in **Figure 5** with V_p in dry rock increasing by about 2.0 km/s and V_s by 1.5 km/s as P_{Diff} varies from almost zero to 50 MPa. Two more examples are shown in

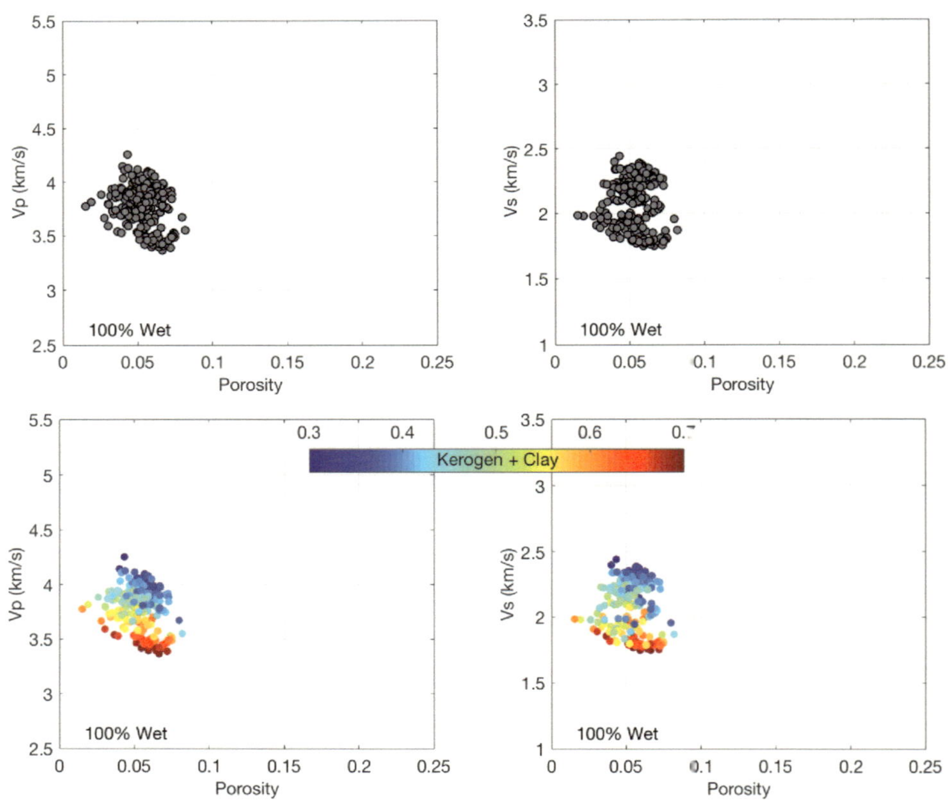

Figure 7.
100% wet rock V_p *(left) and* V_s *(right) without accounting for mineralogy (top) and color-coded by the sum of clay and kerogen contents (bottom) (adopted from Dvorkin et al. [14]).*

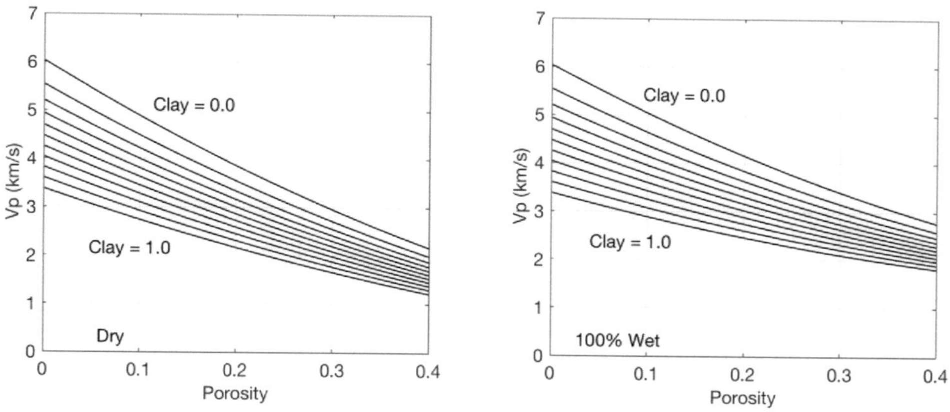

Figure 8.
V_P versus porosity according to the Raymer et al. [3] model for dry rock (left) and 100% water-saturated rock (right). The mineralogy is quartz and clay. The upper curves is for zero clay content, while the bottom curve is for 100% clay. The in-between curves are for gradually increasing clay content with increment 10% (top to bottom).

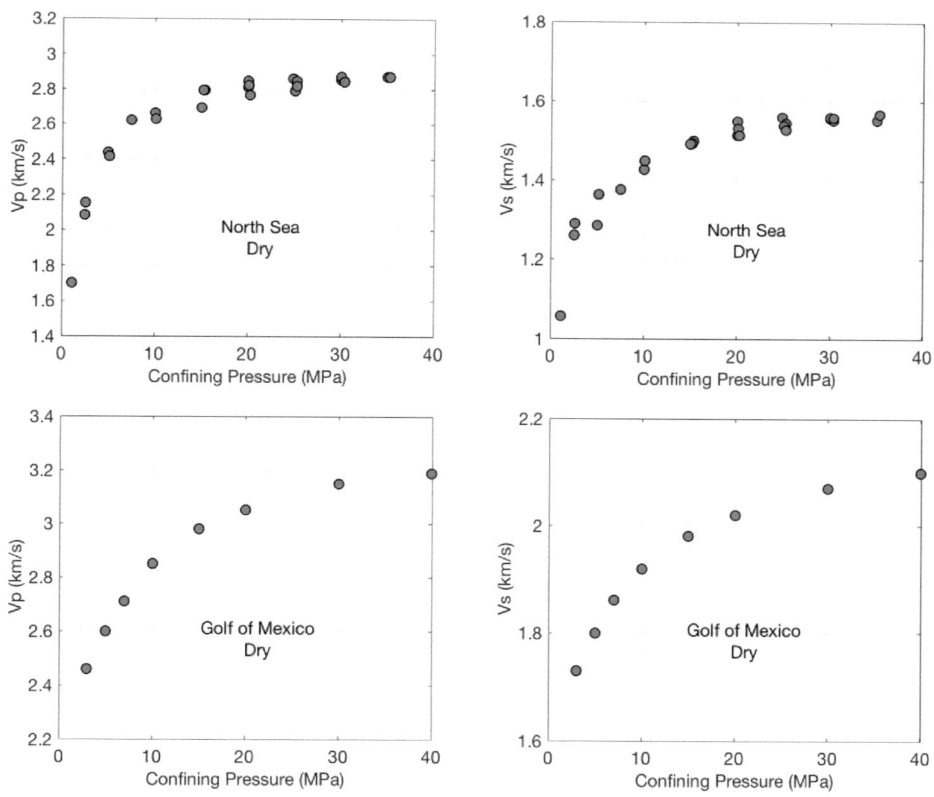

Figure 9.
V_P (left) and V_s (right) versus pressure for two dry sandstone samples from the North Sea (top) and Gulf of Mexico (bottom).

Figure 9, one for a sandstone from the North Sea and the other from the Gulf of Mexico.

The velocity in carbonate rocks is often not as affected by stress as it is in clastic samples. The magnitude of this effect is often influenced by the presence of compliant cracks in the rock. Such cracks act to strongly affect the velocity at low

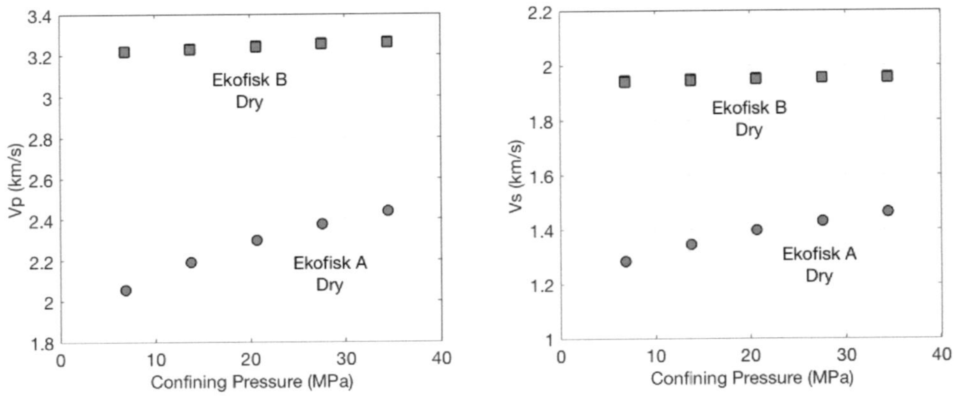

Figure 10.
*Same as **Figure 9** but for two chalk samples from Ekofisk field in the North Sea. Velocities in sample A (porosity 0.38) is pressure dependent, while the velocities in sample B (porosity 0.31) hardly vary with pressure.*

pressure while they are open. As the pressure increases, these cracks close acting to increase the velocity (**Figure 10**, Sample A). In samples where the cracks are absent, the velocity hardly varies as a function of pressure (**Figure 10**, Sample B).

Notice that both historic velocity-porosity model by Wyllie et al. [2] and Raymer et al. [3] do not account for the dependence of the elastic-wave velocities on the confining stress. Both models are suitable for predicting the elastic properties at high, but not at low stress.

The velocity-stress dependence is important in understanding and predicting the seismic responses during hydrocarbon recovery, a process where the differential pressure may increase during production if the reservoir is depleted and the pore pressure is reduced, while the overburden remains constant. This differential pressure may decrease during enhanced oil recovery where water or gas are injected into the reservoir at high pressure, acting to reduce the difference between the overburden and pore pressure. Plots similar to that shown in **Figure 5** (right-hand frame) are useful in simultaneously assessing the effects of the pore fluid and differential pressure on the elastic attributes.

4. Theoretical velocity-porosity models

There are two kinds of elastic moduli versus porosity effective medium models: (a) inclusion models and (b) grain-based models. The first kind models build a rock from the zero-porosity endpoint by placing inclusions into the solid matrix [1]. These models are perhaps relevant to some carbonate rocks where the pores appear as inclusions in calcite or dolomite matrix. The second kind assumes that the rock is formed by solid grains which comprise an uncemented grain pack at the high-porosity endpoint (also called the critical porosity) and, as the porosity is reduced, the original pack is altered either by grain contact cement or by smaller grains deposited in the pore space between the original larger grains, or a combination of these two processes.

As an example of the **inclusion models**, consider the differential effective medium model (DEM), where spheroidal pores are placed inside the solid matrix. A spheroid is an ellipsoid with two large diameters equal to each other and the third diameter smaller or equal to these two. The ratio of the small to large diameter is called the aspect ratio $\alpha \leq 1$. If the spheroid is a sphere, $\alpha = 1$. The inputs are the bulk and shear moduli of the mineral matrix and those of the inclusions.

Figure 11 (top) shows how the bulk and shear moduli depend on the total porosity for pure calcite rock with the bulk and shear moduli of the mineral 76.8 and 32.0 GPa, respectively, and its density 2.71 g/cc. The pores are empty, meaning the bulk and shear moduli of the inclusions are zero. In the same figure (bottom), we plot the respective V_p and V_s. The aspect ratio is different for each of the curves shown. It is 1.00 for the upper curves and gradually decreases to 0.50, 0.20, 0.10, and 0.01 for the curves below. The smaller the aspect ratio, the smaller the elastic moduli and velocities at a fixed porosity.

Figure 12 is the same as **Figure 11** except that we use a single aspect ratio 0.10 and compare the results for empty inclusions with those for water-filled inclusions where the bulk modulus is 2.25 GPa and density is 1.00 g/cc.

We observe that *both* the bulk and shear moduli increase for pores filled with water as compared to empty pores. So do V_p and V_s. This means that DEM is not consistent with Gassmann's fluid substitution theory [9] which predicts that the shear modulus is pore-fluid-independent and V_s reduces upon saturation due to increasing bulk density.

Notice that DEM curves connect two endpoints, one at zero porosity where the elastic moduli of rock are those of the mineral matrix and the other at 100% porosity where the elastic moduli are those of the inclusions (fluid in the pores). About three decades ago, Nur observed that most natural rocks simply do not exist in the entire zero to 100% porosity range. The maximum geologically plausible porosity for clastic rocks (sands and sandstones) is about 0.40. It may be higher in

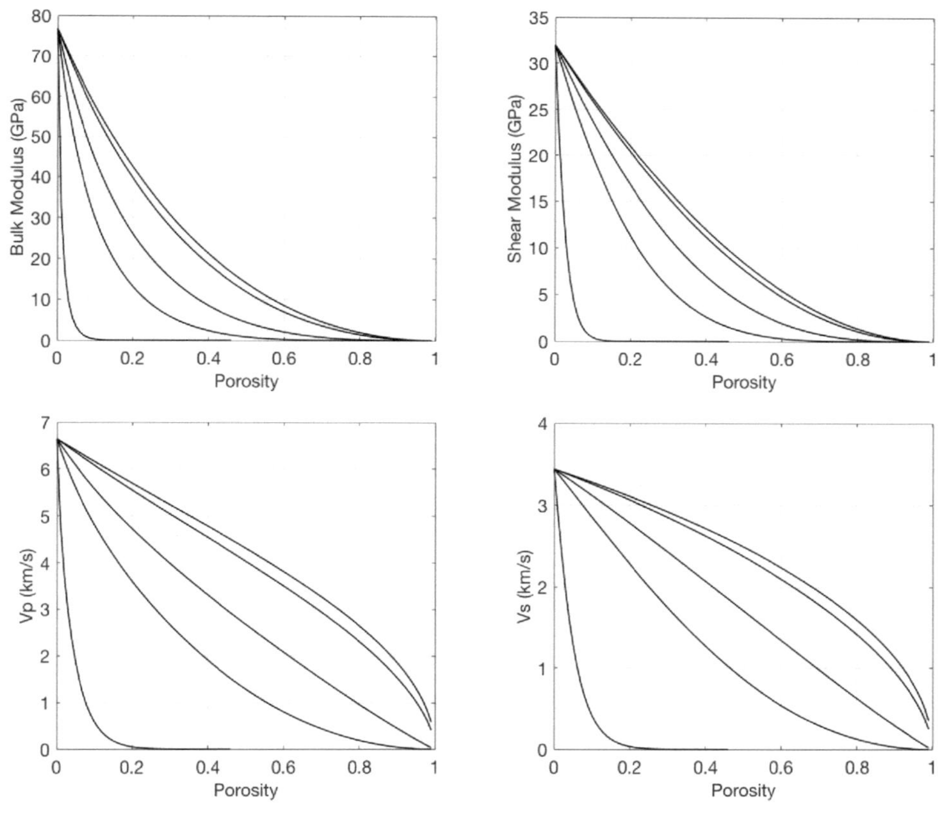

Figure 11.
Elastic moduli (top) and velocities (bottom) versus porosity computed using DEM model for a pure calcite rock. The aspect ratio corresponding to the top curves is 1.00 and for the bottom curve 0.01. The aspect ratio gradually decreases to 0.50, 0.20, and 0.10 for the curves in between (top to bottom).

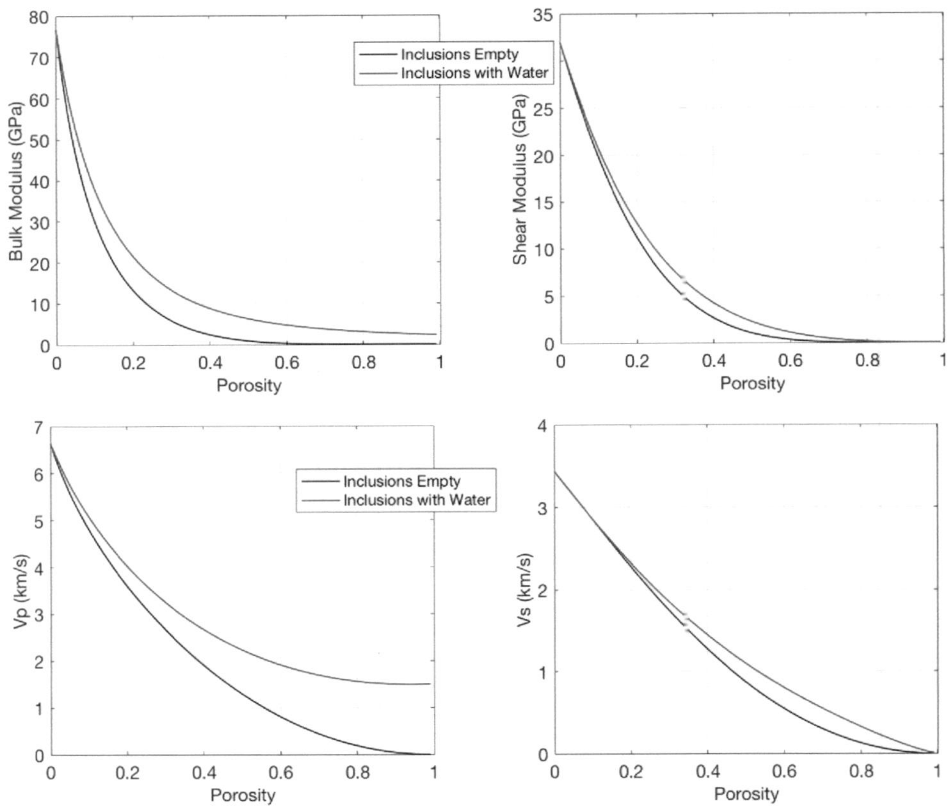

Figure 12.
*Same as **Figure 11** but for a single aspect ratio 0.10 and for empty pores (black) and pores filled with water (blue).*

carbonates, such as chalks, that can have porosity up to 0.50. This porosity can be even higher for foam-like formations, such as volcanic rock (pumice) or artificially manufactured glass foam. This maximum porosity is called the **critical porosity**. This concept was formalized in Nur et al. [15].

One implication of the critical porosity concept is that the high-porosity end-point should be at the critical porosity rather than at 100% porosity. It gave rise to the so-called modified elastic bounds. The simplest example is based on the upper elastic bound (also called the Voigt bound) for a composite made of two elastic components ("1" and "2") with the compressional and shear moduli M_1, G_1 and M_2, G_2, respectively.

Assume that $M_2 = G_2 = 0$. Then the respective moduli of this composite (M and G) at porosity ϕ cannot exceed

$$M = (1 - \phi)M_1; G = (1 - \phi)G_1. \tag{29}$$

These two curves are plotted in **Figure 13**. In the same figure, we plot Han's [13] data for low-clay-content samples at 50 MPa confining pressure. These data fall way below the upper bound curves for pure quartz with $M_1 = 96.6$ GPa and $G_1 = 45.0$ GPa.

The modified bounds use the same equations, but with porosity scaled by the critical porosity ϕ_c:

$$M = (1 - \phi/\phi_c)M_1; G = (1 - \phi/\phi_c)G_1; \phi \leq \phi_c, \tag{30}$$

giving modified curves that are much closer to the data (**Figure 13**).

All **grain-based theories** exploit the critical porosity concept. We start with the **contact-cement** theory where it is assumed that the grains are not subjected to any confining stress at ϕ_c, and, as a result, the elastic moduli are zero, and porosity reduction is due to cement rims enveloping the grains (**Figure 14**). Such contact cement acts to rapidly increase the elastic moduli of the grain pack due to the dramatically expanding contact areas between the grains as porosity decreases, as explained in Dvorkin et al. [4], where the theoretical equations are given as well. This model is only valid in the very high-porosity range.

The **soft-sand** model assumes that at the critical porosity and the elastic properties of the grain pack are given by the Hertz-Mindlin [16] contact theory. This theory assumes that the grain pack is made of identical spherical grains whose elastic properties are those of the mineral (solid) matrix as given by Eq. (25). Combined with the mean field approximation that assumes that all grains are subject to identical local stresses and have the same average number of contacts per

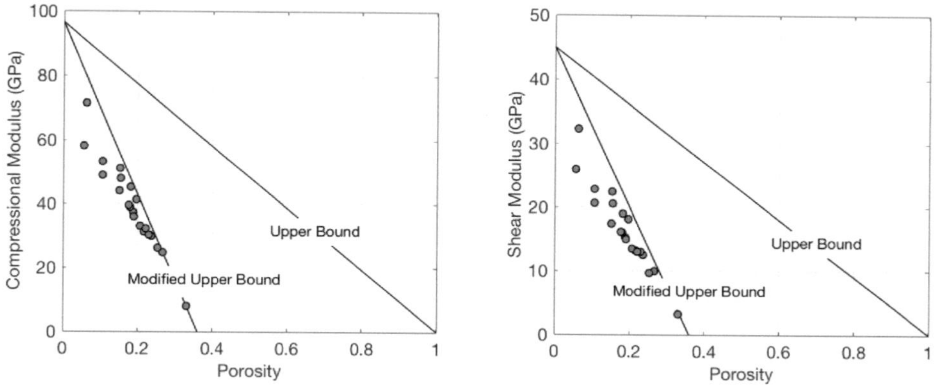

Figure 13.
Upper and modified upper elastic bounds for the compressional (left) and shear (right) moduli versus porosity. The critical porosity is 0.36. Data are from Han's [13] sandstone dataset for the clay content below 7% and with the elastic-wave velocities measured on dry samples at 50 MPa confining pressure.

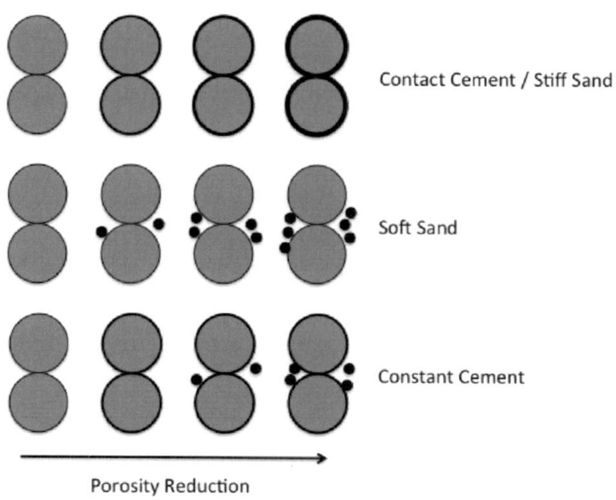

Figure 14.
Schematic modes of porosity reduction. From top to bottom: Contact-cement and stiff-sand model; soft-sand model; and constant-cement model (adopted from Dvorkin et al. [4]).

grain n (also called the coordination number), the respective dry rock bulk (K_{HM}) and shear (G_{HM}) moduli are

$$K_{HM} = \left[\frac{n^2(1-\phi_c)^2 G_s^2}{18\pi^2(1-\nu_s)^2} P\right]^{\frac{1}{3}}, \quad G_{HM} = \frac{5-4\nu_s}{5(2-\nu_s)}\left[\frac{3n^2(1-\phi_c)^2 G_s^2}{2\pi^2(1-\nu_s)^2} P\right]^{\frac{1}{3}}, \tag{31}$$

where P is the differential pressure (Eq. (28)) and G_s and ν_s are the shear modulus and Poisson's ratio of the solid matrix, respectively. This model implies that porosity reduction is not due to contact-cement deposition but instead due to smaller particles deposited away from grain contacts (**Figure 14**).

The coordination number n in an identical grain pack at the critical porosity is about 6.

It is assumed in Eq. (31) that the grains have infinite friction (no slip) at their contacts. If we allow only the fraction f of these contacts to have infinite friction while the rest of the contacts are frictionless and can slip, the equation for K_{HM} does not change but G_{HM} becomes now

$$G_{HM} = \frac{2+3f-\nu(1+3f)}{5(2-\nu)}\left[\frac{3n^2(1-\phi_c)^2 G^2}{2\pi^2(1-\nu)^2} P\right]^{\frac{1}{3}}. \tag{32}$$

This parameter f is called the shear stiffness correction factor.

Finally, to obtain the dry rock bulk (K_{Soft}) and shear (G_{Soft}) moduli at any porosity $\phi < \phi_c$, we use the modified (critical porosity scaled) lower Hashin-Shtrikman bound (e.g., [4]):

$$\begin{aligned} K_{Soft} &= \left[\frac{\phi/\phi_c}{K_{HM}+\frac{4}{3}G_{HM}} + \frac{1-\phi/\phi_c}{K+\frac{4}{3}G_{HM}}\right]^{-1} - \frac{4}{3}G_{HM}, \\ G_{Soft} &= \left[\frac{\phi/\phi_c}{G_{HM}+z_{HM}} + \frac{1-\phi/\phi_c}{G+z_{HM}}\right]^{-1} - z_{HM}, \quad z_{HM} = \frac{G_{HM}}{6}\left(\frac{9K_{HM}+8G_{HM}}{K_{HM}+2G_{HM}}\right). \end{aligned} \tag{33}$$

It is important to emphasize that the critical porosity endpoints here do not necessarily have to be given by the Hertz-Mindlin contact theory. Alternatively, these values can be selected from experimental data. What is most important in this model is the usage of the "soft" connection between the two porosity endpoints.

An alternative "stiff" connection between the aforementioned endpoints is given by the modified upper Hashin-Shtrikman bound as

$$\begin{aligned} K_{Stiff} &= \left[\frac{\phi/\phi_c}{K_{HM}+\frac{4}{3}G_s} + \frac{1-\phi/\phi_c}{K+\frac{4}{3}G_s}\right]^{-1} - \frac{4}{3}G_s, \\ G_{Stiff} &= \left[\frac{\phi/\phi_c}{G_{HM}+z} + \frac{1-\phi/\phi_c}{G_s+z}\right]^{-1} - z, \quad z = \frac{G_s}{6}\left(\frac{9K_s+8G_s}{K_s+2G_s}\right), \end{aligned} \tag{34}$$

where, once again, G_s and K_s are the shear and bulk moduli of the solid matrix, respectively.

This stiff connection, also called the stiff-sand model, can serve to connect the contact-cement curve with the zero-porosity endpoint.

Yet another model belonging to this family is the constant-cement model. It assumes that the grains have initial contact cementation with further porosity reduction due to the placement of small particles away from grain contacts (**Figure 14**). The functional form of this model is the same as in the soft-sand model (Eq. (33)) but with artificially high coordination number.

Examples of velocity-porosity curves according to the aforementioned grain-based theories are shown in **Figure 15**, where we assumed that both the grain and cement materials are pure quartz; n for the soft-sand model is 6, while it is 20 for the constant-cement model; and the differential pressure is 20 MPa. The shear stiffness correction factor is 1.

Figure 16 shows an example of using the constant-cement model to describe the elastic behavior of unconventional gas shale, while **Figure 17** is an example of applying the stiff-sand model to carbonate reservoirs. The parameters of the models are provided in the captions. These two examples show that the grain-based theories given here are appropriate not only for clastic sediments but also in very different lithological settings.

Figure 18 shows laboratory data obtained at 30 MPa confining pressure on dry high-porosity, almost pure-quartz sand samples from the North Sea. In this classic example, the higher-velocity dataset is contact-cemented turbidite sand, while the

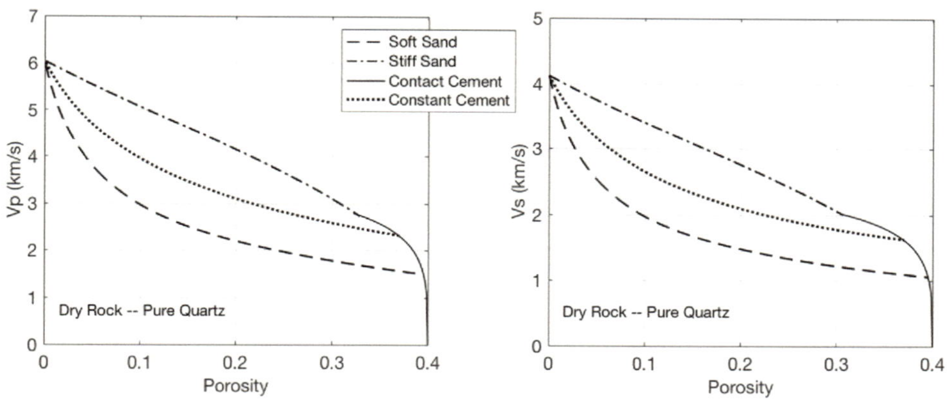

Figure 15.
Velocity-porosity curves according to the soft-sand, stiff-sand, contact-cement, and constant-cement models as explained in the text.

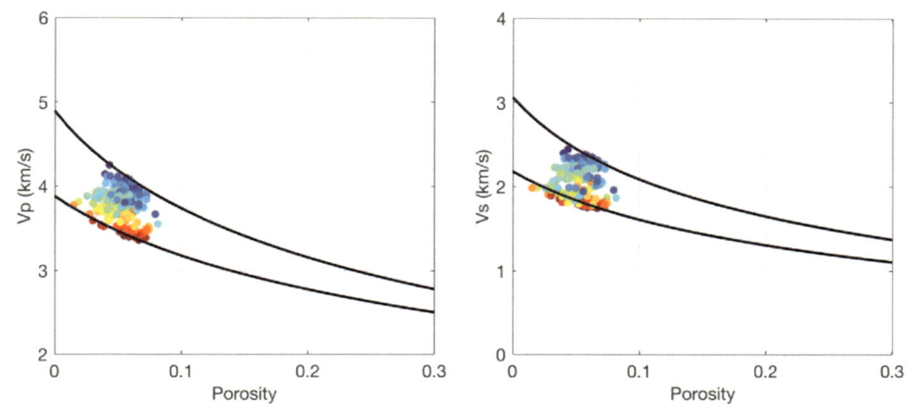

Figure 16.
V_p *(left) and* V_s *(right) versus porosity for gas shale from wireline data adjusted for 100% water saturation. The color code is the sum of the clay and kerogen volume fractions (red for high and blue for low). The model curves are computed to bound the data. These curves are from the constant-cement model with the coordination number 12, differential pressure 26 MPa, critical porosity 0.40, and shear stiffness correction factor 1 (adopted from Dvorkin et al. [14]).*

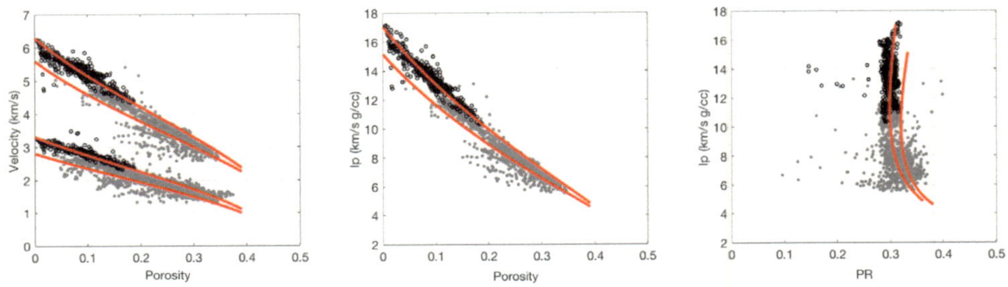

Figure 17.
Velocity- (left) and impedance-porosity (middle) plots showing chalk (gray) and lower-porosity carbonate (black) data points from wireline data adjusted for 100% water saturation. Graph on the left is the impedance versus Poison's ratio plot, also for 100% water saturation conditions. The curves are from the stiff-sand model with the coordination number 6, differential pressure 30 MPa, critical porosity 0.40, and shear stiffness correction factor 1. The two model curves are for the two slightly different properties of the pure calcite end member (adopted from Dvorkin and Alabbad [17]).

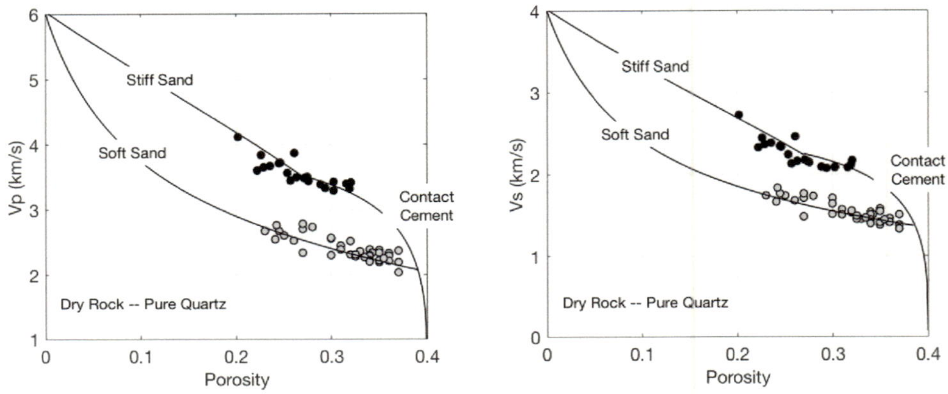

Figure 18.
V_P *(left) and* V_s *(right) versus porosity for two high-porosity sand datasets as explained in the text. The model curves marked in the plots are computed for 30 MPa differential pressure, critical porosity 0.40, coordination number 7, shear stiffness correction factor 1, and dry rock.*

lower-velocity dataset is friable and virtually uncemented sand. The former data can be matched by the contact-cement curves transitioning into the stiff-sand trajectories. The latter data are matched by the soft-sand curves.

5. Digital rock physics

Digital rock physics is based on the concept "image and compute," image rock at the pore scale (**Figure 19**) and then simulate in the computer various processes in such an image to arrive at a desired rock property. These simulations include viscous fluid flow to arrive at hydraulic permeability, electrical charge flow to arrive at electrical resistivity, as well as elastic deformation to arrive at the elastic moduli and velocities.

The advantage of such digital approach is that the same sample can be reused multiple times, unlike in physical experiments where a sample is altered after every test; the sample can be digitally altered by, e.g., introducing digenetic cementation, which is hardly possible in physical experiments, as well as subsampling of a digital volume to investigate how various rock properties vary within the volume and how relations between rock properties depend on the spatial scale of investigation.

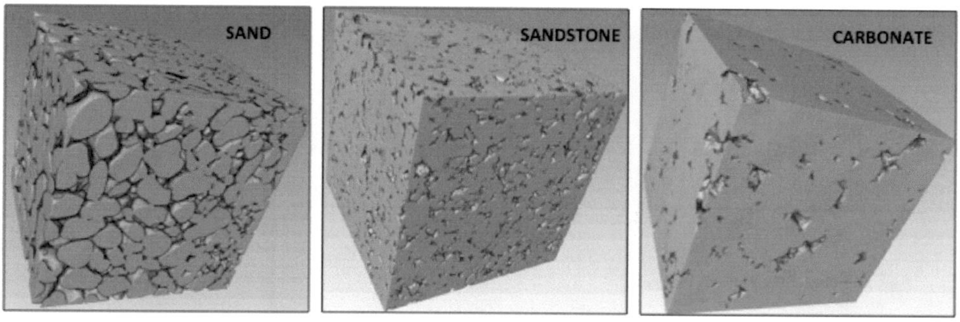

Figure 19.
Segmented digital images of loose sand (porosity about 30%), sandstone (porosity about 20%), and carbonate (porosity about 15%) showing the mineral matrix and pores. The images are a few mm across.

Although the aforementioned concept is simple, its implementation is not. First, the imaging has to be conducted at the appropriate scale and resolution to reveal the salient features of natural rock relevant to the process under examination. Second, the image has to be segmented to separate minerals from pores and segregate various minerals within the solid matrix, as well as fluid phases inside the pores. Third, powerful computational engines have to be utilized and verified to simulate processes relevant to the physical experiment.

In spite of these complexities, during the last decade, DRP has emerged as a powerful technique complementing (if not replacing) physical testing, mostly due to the recent advances in imaging hardware and image processing and computational software, the latter combined with steadily improving computational power. Not only DRP has become a novel research tool in academia and national labs, but is has also been adopted by leading oil and service companies.

There is one more inherent feature of DRP that needs to be accounted for. Pore-scale rock images are only a few mm in size, and the higher the resolution needed to revel the salient features, the smaller the field of view. At the same time, these computational results have to be relevant at much larger spatial scales of feet for wireline measurement interpretations in the well or tens and hundreds of feet in seismic prospecting. Even such basic property as porosity may be different if measured on an inch-sized sample an on mm-sized fragment of the same sample.

One way out of this conundrum is instead of directly comparing data points generated by different methods of measurement, compare trends formed by such data points, such as permeability versus porosity trends. Dvorkin et al. [18] show that such trends are often hidden inside a very small digital sample and can be derived by subsampling it. Moreover, these computational trends often match

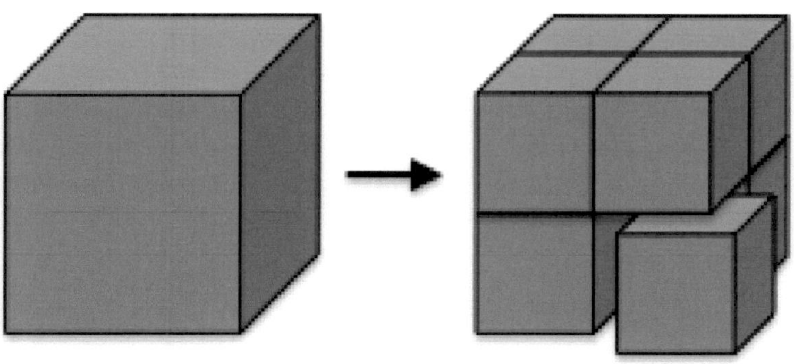

Figure 20.
Illustration of the subsampling approach.

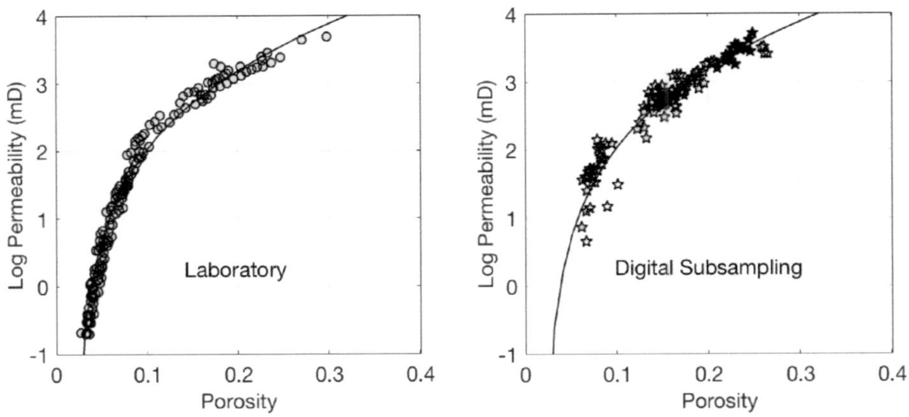

Figure 21.
Permeability versus porosity in Fontainebleau sandstone. Left: Laboratory data matched with a Kozeny-Carman theoretical curve. Right: Multiple permeability versus porosity data points computed from a few digital Fontainebleau samples and subsamples thereof (adopted from Dvorkin et al. [18]).

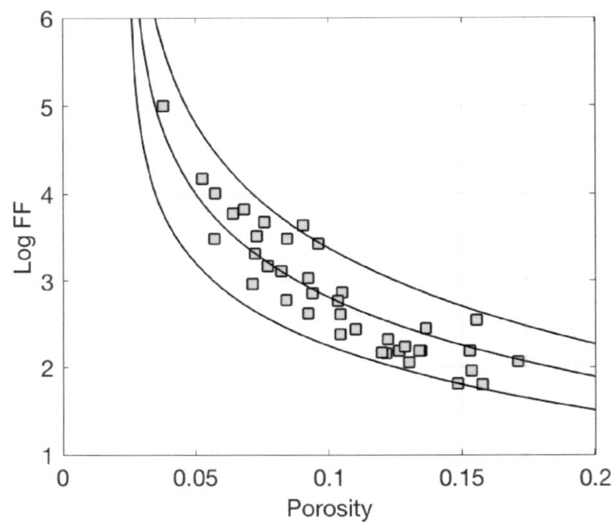

Figure 22.
Formation factor versus porosity computed on carbonate cuttings. The curves are from Archie's equation with the cementation exponent m 2.0, 2.5, and 3.0 (bottom to top) (adopted from Dvorkin et al. [18]).

relevant physical trends and/or theoretical rock physics transforms, hence validating computational results and making them relevant at much coarser spatial scales.

The approach is to subsample a digital volume into 2^3, 3^3, or 4^3 subvolumes (**Figure 20**) and then compute the desired property pairs (e.g., porosity and permeability) on each of these subvolumes. Very often, the property pairs thus computed form a meaningful trend supported by physical measurements and/or theories (see examples in **Figures 21–23**). We can call this subsampling approach "to see the rock in a grain of sand."

These results open ways to a meaningful utilization of DRP in research and industry. Publications related to DRP are many and the number is growing. We refer the reader to Kameda and Dvorkin [19], Dvorkin et al. [20], Dvorkin and Derzhi [21], and Andra et al. [22, 23].

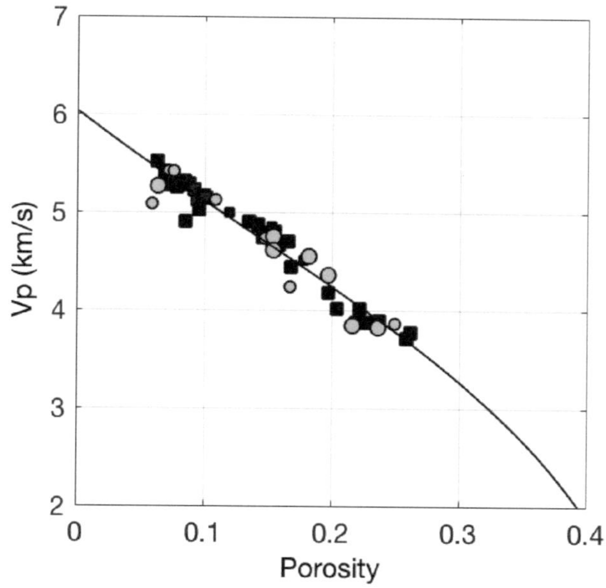

Figure 23.
V_P *versus porosity for Fontainebleau sandstone as computed from a few digital samples and subsamples thereof (squares). Gray circles are from laboratory measurements of dry samples. The curve is from the stiff-sand model (adopted from Dvorkin et al. [18]).*

6. Conclusion

This chapter presents an overview of rock physics, starting with its history and ending with the most recent development, the digital rock physics. This chapter can be used as a basic reference pointing towards published sources where the topic is developed in-depth and detailed equations, tables, and experimental results are given. One of such comprehensive sources is the third edition of the *Rock Physics Handbook* [24].

Rock physics remains a key component in interpreting seismic and other remote sensing data for the underlying properties and conditions of the subsurface. A plethora of such practical results has appeared and continues to appear in geophysical journals, such as *Geophysics* (Society of Exploration Geophysicists), *Journal of Geophysical Research* (American Geophysical Union), and *First Break* (European Association of Geoscientists and Engineers), as well as presented at conferences worldwide.

An important topic not addressed in this chapter is a simultaneous interpretation of different remote sensing sources, such as seismic prospecting, electric and electromagnetic sensing, and gravity methods. Once again, such materials can be found in the proceedings and books from the aforementioned professional societies.

We feel that the material presented can serve as a detailed introduction into the extensive field of physics of rocks and be of use to graduate students, as well as advanced professional in earth and environmental sciences.

Author details

Jack Dvorkin
College of Petroleum and Geosciences (CPG), King Fahd University of Petroleum and Minerals (KFUPM), Dhahran, Saudi Arabia

*Address all correspondence to: jackdvorkin007@gmail.com

References

[1] Mavko G, Mukerji T, Dvorkin J. Rock Physics Handbook. 2nd ed. Cambridge, UK: Cambridge University Press; 2009

[2] Wyllie MRJ, Gregory AR, Gardner LW. Elastic wave velocities in heterogeneous and porous media. Geophysics. 1956;**21**:41-70

[3] Raymer LL, Hunt ER, Gardner JS. An improved sonic transit time-to-porosity transform. In: Transactions of the Society of Professional Well Log Analysts, 21st Annual Logging Symposium, Paper P, Houston, TX, USA. 1980

[4] Dvorkin J, Gutierrez M, Grana D. Seismic Reflections of Rock Properties. Cambridge, UK: Cambridge University Press; 2014

[5] Gomez CT. Reservoir characterization combining elastic velocities and electrical resistivity measurements [PhD thesis]. Stanford, CA, USA: Stanford University. 2009

[6] Archie GE. The electrical resistivity log as an aid in determining some reservoir characteristics. Transactions of the American Society of Mechanical Engineers. 1942;**146**:54-62

[7] Carman PC. Lecoulement des Gaz a Travers les Milieux Poreux, Paris: Bibliotheque des Sciences et Techniques Nucleaires, Presses Universitaires de France; 1961

[8] Zimmer MA. Seismic velocities in unconsolidated sands: Measurements of pressure, sorting, and compaction effects [PhD thesis]. Stanford, CA, USA: Stanford University; 2003

[9] Gassmann F. Elasticity of porous media: Uber die elastizitat poroser medien. Vierteljahrsschrift der Naturforschenden Gesselschaft. 1951; **96**:1-23

[10] Mavko G, Chan C, Mukerji T. Fluid substitution: Estimating changes in V_p without knowing V_s. Geophysics. 1995; **60**:1750-1755

[11] Muqtadir A, Al-Dughaimi S, Dvorkin JP. Deformation of granular aggregates: Static and dynamic bulk moduli. JGR Solid Earth. 2020;**125**:1-12

[12] Batzle M, Wang Z. Seismic properties of pore fluids. Geophysics. 1992;**57**:1396-1408. DOI: 10.1190/1.1443207

[13] Han D-H. Effects of porosity and clay content on acoustic properties of sandstones and unconsolidated sediments [PhD thesis]. Stanford, CA, USA: Stanford University; 1986

[14] Dvorkin J, Walls J, Davalos G. Rock physics of unconventional shale: Wireline data and theory. In: 5th International Rock Physics Workshop, Milan, 2020. 2020

[15] Nur A, Mavko G, Dvorkin J, Galmudi D. Critical porosity: A key to relating physical properties to porosity in rocks. The Leading Edge. 1998;**17**: 357-362

[16] Mindlin RD. Compliance of elastic bodies in contact. Transactions ASME. 1949;**71**:A-259

[17] Dvorkin J, Alabbad A. Velocity-porosity-mineralogy trends in chalk and consolidated carbonate rocks. Geophysical Journal International. 2019; **219**:662-671

[18] Dvorkin J, Derzhi N, Diaz E, Fang Q. Relevance of computational rock physics. Geophysics. 2011;**76**:E141-E153

[19] Kameda A, Dvorkin J. To see a rock in a grain of sand. The Leading Edge. 2004;**23**:790-794

[20] Dvorkin J, Fang Q, Derzhi N. Etudes in computational rock physics: Alterations and benchmarking. Geophysics. 2012;77:D45-D52

[21] Dvorkin J, Derzhi N. Rules of upscaling for rock physics transforms: Composites of randomly and independently drawn elements. Geophysics. 2012;77:WA129-WA139

[22] Andra H, Combaret N, Dvorkin J, Glatt E, Han J, Kabel M, et al. Digital Rock Physics Benchmarks—Part I: Imaging and Segmentation, Computers and Geosciences. Amsterdam, The Netherlands: Elsevier; 2013. pp. 1-8

[23] Andra H, Combaret N, Dvorkin J, Glatt E, Han J, Kabel M, et al. Digital Rock Physics Benchmarks—Part II: Computing Effective Properties, Computers and Geosciences. Amsterdam, The Netherlands: Elsevier; 2013. pp. 1-11

[24] Mavko G, Mukerji T, Dvorkin J. Rock Physics Handbook. 3rd ed. Cambridge, UK: Cambridge University Press; 2020

Chapter 4

Enhanced Oil Recovery: Chemical Flooding

Ahmed Ragab and Eman M. Mansour

Abstract

The enhanced oil recovery phase of oil reservoirs production usually comes after the water/gas injection (secondary recovery) phase. The main objective of EOR application is to mobilize the remaining oil through enhancing the oil displacement and volumetric sweep efficiency. The oil displacement efficiency enhances by reducing the oil viscosity and/or by reducing the interfacial tension, while the volumetric sweep efficiency improves by developing a favorable mobility ratio between the displacing fluid and the remaining oil. It is important to identify remaining oil and the production mechanisms that are necessary to improve oil recovery prior to implementing an EOR phase. Chemical enhanced oil recovery is one of the major EOR methods that reduces the residual oil saturation by lowering water-oil interfacial tension (surfactant/alkaline) and increases the volumetric sweep efficiency by reducing the water-oil mobility ratio (polymer). In this chapter, the basic mechanisms of different chemical methods have been discussed including the interactions of different chemicals with the reservoir rocks and fluids. In addition, an up-to-date status of chemical flooding at the laboratory scale, pilot projects and field applications have been reported.

Keywords: enhanced oil recovery, EOR, chemical flooding, alkaline, surfactant, polymer, interfacial tension, IFT

1. Introduction

The Average oil recovery after the primary recovery phase is about 5–20% of the original oil in place (OOIP) and can be increased by applying the secondary recovery phase up to 40%. Usually, the EOR application stage will be after the secondary recovery when the main challenge is not the reservoir pressure only, but also the reservoir fluids relative mobility compared to the injected fluids during the secondary recovery phase [1].

There are different EOR methods such as thermal recovery, miscible Gas Injection, Chemical flooding and Microbial EOR as shown in **Figure 1**. This chapter covers the fundamentals and the mechanisms of the recovery enhancement of the chemical flooding EOR as one of the main EOR methods [2]. The feasibility study and design for EOR projects require integrated work between different disciplines such as reservoir engineers, petroleum geologists, petrophysits, geomodellers, chemical engineers, and production engineers whom are responsible to start with the screening phase of the different EOR methods and come up with the shortlisted one in order to go for the next step which is lab testing phase that requires PVT/core labs capable to implement the various EOR lab tests, then, analyze the lab scale results to be coupled with the reservoir simulation model in order to estimate the incremental recovery for the different EOR methods under study. For any EOR project, the initial stage is the screening criteria in order to identify the best EOR application for the candidate reservoirs in terms of incremental recovery that will be added and the economics of the project [3]. For any EOR project, the initial stage is the screening criteria in order to identify the best EOR application for the candidate reservoirs in terms of incremental recovery that will be added and the economics of the project. The screening criteria is based on both reservoir rock and fluids properties such as oil gravity, oil viscosity, oil composition, remaining oil saturation (target), formation type, reservoir thickness, depth, and temperature. In **Table 1**, a summary of screening criteria for the chemical EOR methods based on lab and applied field data. So, in this chapter we are assuming that the screening criteria was done and it has been found that the chemical flooding is the optimum EOR method that can be applied for the reservoir under study [4].

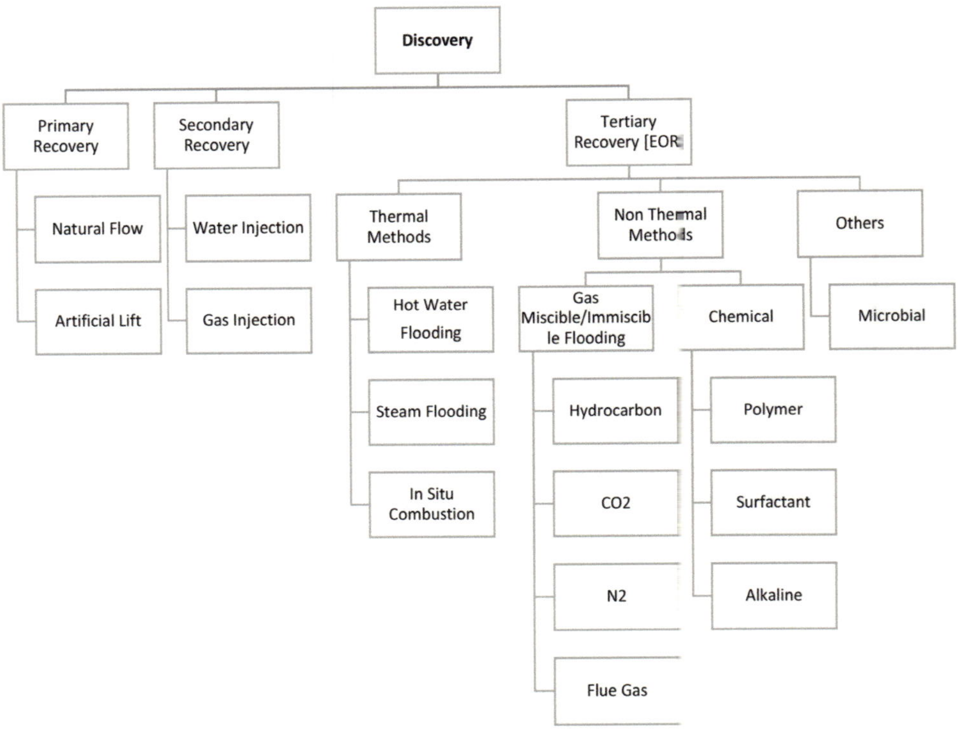

Figure 1.
Oil production mechanisms.

EOR Method	Oil Gravity (°API)	Viscosity (cP)	Oil Saturation (%)	Formation Type	Reservoir Thickness (ft)	Average Permeability (mD)	Depth (ft)	Temperature (°F)
Polymer Flooding	> 15	<150 & >10	>50	Sandstone is preferred	NC	>10	<9000	>200
Micellar/Polymer, ASP, and Alkaline flooding	> 20	<35	>35	Sandstone is preferred	NC	>10	>9000	>200

Table 1.
A summary of screening criteria for the chemical EOR methods.

2. Chemical EOR flooding classification

The chemical flooding EOR can be categorized into the following:

1. Polymer flooding.

2. Surfactant flooding.

3. Surfactant-polymer flooding.

4. Alkaline flooding.

5. Alkaline-surfactant-polymer flooding.

The next sections of this chapter will discuss individually each method in order to illustrate the fundamentals, the reservoir-fluids interactions processes, and the field applications [5].

2.1 Polymer flooding

In concept, a water-soluble polymer is used to reduce the mobility ratio of water-oil by increasing the water viscosity which improves the volumetric sweep efficiency. The mechanism of polymer flooding is to increase the water viscosity and also to reduce the permeability of the rock to water, in other words, to reduce the water-oil mobility ratio close to unity or less [6]. Over the past years, polymer floods projects have been applied over a wide range of conditions:

- Reservoir temperatures [46–235]°F.

- Average reservoir permeability [0.6–15,000] mD.

- Oil viscosity [0.01–1494] cP.

- Net pay thickness [4–432] ft.

- Remaining oil at start-up [36–97] % of OOIP.

Polymers have been used in oil production in three modes; as near-well treatments to improve the performance of water injectors or watered out producers by blocking off high conductivity zones, agents that may be cross-linked in situ to plug high conductivity zones at depth in the reservoir and agents to lower water mobility or water-oil mobility ratio. Polymer flooding is suited for reservoirs where normal water floods fail due to one of the two reasons: High Heterogeneity and High oil water mobility ratio which is targeting the oil in areas of the reservoir that have not been contacted efficiently [7]. The main economic limitation is the cost of polymers is. For example, if the cost of acrylamide/acrylate copolymers and xanthan polymers were substantially lower and higher polymer concentrations with larger polymer flood bank sizes could be granted in a given application. It would improve oil-recovery efficiencies, wider ranges of potential applications, and higher profits. Polymer flooding is showing promising results, specifically, if flooding projects are started at high remaining oil saturations. Polymer flooding has been conducted in sandstone and carbonate reservoirs, and oil-wet, water-wet, and mixed-wettability reservoirs [8].

2.1.1 Mechanism

The main effect of the polymer is the enhancement of the water-oil mobility ratio to be unity or less, the mobility ratio is defined as the ratio of the mobility of displacing phase to the mobility of displaced phase which is calculated from the following equation [6].

$$M_{w-o} = \frac{M_w}{M_o} = \frac{K_w}{K_o} * \frac{\mu_o}{\mu_w} = \frac{K_{rw}}{K_{ro}} * \frac{\mu_o}{\mu_w} \quad (1)$$

where
M_{w-o}:the water – oil mobility ratio
M_w:the water mobility
M_o:the oil mobility
K_w:the effective permeability to water, mD
K_o:the effective mobility to oil, mD
μ_o:the oil viscosity, cP
μ_w:the water viscosity, cP
K_{rw}:the relative permeability to water
K_{ro}:the relative permeability to water

As per this equation, it is clear that in order to drive the mobility ration to be unity or less, the water viscosity is increased by adding the water-soluble polymers to the injected water as shown in **Figure 2**, when the displacing fluid (water) viscosity is lower than the oil, the recovery efficiency decreases as the remaining oil after this flooding is about 45% of the OOIP at 0.1 viscosity ratio. On the other hand, once the viscosity ration reached to 1 (polymer added to water) the remaining oil after the flooding will be reduced to 20% of the OOIP. As summary, the highest viscosity ratio is the highest oil recovery [9].

2.1.2 Polymer flooding advantages

The advantages of polymer flooding could be summarized as following [10]:

I. Applicable over a wide range of conditions.

II. A reduction in the quantity of water required to reduce the oil saturation to its residual value in the swept portion of the reservoir.

III. An increase in the areal and vertical coverage in the reservoir due to a reduced water flood mobility ratio.

IV. Diverting the injected from swept zones.

V. Promising for heavy oil application.

VI. Cost-effective.

Polymer flooding limitations are:

I. High oil viscosities require a higher polymer concentration.

II. Results are normally better if the polymer flood is started before the water-oil ratio becomes excessively high.

III. Clays increase polymer adsorption.

IV. Some heterogeneity are acceptable, but avoid extensive fractures.

V. Lower injectivity than with water can adversely affect oil production rates in the early stages of the polymer flood.

VI. Xanthan gum polymers cost more, are subject to microbial degradation, and have a greater potential for wellbore plugging.

2.1.3 Field projects statistical data of some polymer flooding

A summary statistical data for field projects of polymer flooding as shown in **Table 2**.

2.2 Surfactant flooding

Correctly designed surfactants can create micro emulsions at the interface between oil and water phases, which cause a reduction in the interfacial tension (IFT) that consequently will mobilize the residual oil which improving the oil recovery as shown in **Figure 3**. This method of EOR is a challenging one by many factors such as rock adsorption of the surfactant and co-surfactant, and the chromatographic separation of the surfactant during the injection in the reservoir. The designed surfactants should be resistant and active at reservoir conditions which could by at higher pressure, temperature and water salinities [11]. In the surfactant flooding the phase behavior is the most important factor to make it successful. Currently, there is no EOS model to describe the phase behavior in these systems. Consequently, phase behavior studies should be observed experimentally which is challenging to mimic the reservoir conditions. Surfactants solutions are used to reduce the oil-water IFT, while the co-surfactants are mixed with these solutions in order to enhance the properties of the surfactant solutions. The co-surfactants added to the solutions are serving as an active agent or a promoter

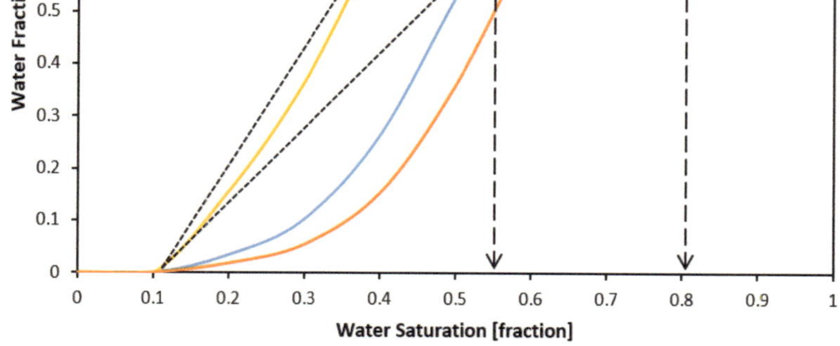

Figure 2.
Effect of viscosity ratio on the fractional flow curve.

Parameter(s)	No. of Projects	Mean
Depth, ft	87	4000
Temperature, F	88	117
Permeability, mD	80	453
Oil Viscosity, cP	82	21.5
Polymer Concentration, ppm	48	279
Oil Recovery, %OOIP	20	4

Table 2.
A summary of statistical data for field projects.

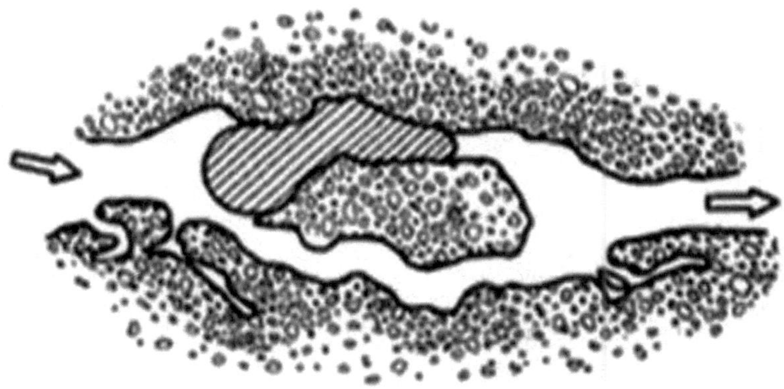

Figure 3.
Principle of flooding, where residual oil is trapped in the reservoir, for the movement of oil through the narrow capillary pores, very low oil/water interfacial tension (IFT) is required.

in the mixed solution in order to enhance the surfactant effectiveness with respect to temperature and water salinity as it is well known that surfactant flooding is sensitive to reservoir temperature and salinity [6].

2.2.1 Mechanism

A surfactant is added to an aqueous fluid and co-surfactant is also added in order to prepare the surfactant solution and injected into the reservoirs as surfactant flooding reduces the interfacial tension between the oil and water phases and also alters the wettability of the reservoir rock in order to mobilize the residual oil trapped in the reservoir which improves the oil recovery as shown in **Figure 3** [12].

The surfactant selection is a critical stage in designing the surfactant flooding projects as the Anionic surfactants preferred due to the following reasons [13]:

- Low adsorption at neutral to high pH on both sandstones and carbonates.

- Can be tailored to a wide range of conditions.

- Widely available at low cost in special cases.
- Sulfates for low temperature applications.
- Sulfonates for high temperature applications.
- Cationic scan be used as co-surfactants.

On the other hand, the Non-ionic surfactants have not performed as well for EOR as anionic surfactants. Sulfonated hydrocarbons such as alcohol propoxylate sulfate or alcohol propoxylate sulfonate are commonly used for Surfactant flooding.

2.2.2 Surfactant flooding advantages

The surfactant flooding has several advantages and some of them are listed below [5]:

1. Very effective in lab test [high oil recovery].
2. Surfactant modeling is relatively simple with only a few well-designed experiments needed to provide the most important simulation parameters.
3. Current high-performance surfactants cost less than $2/lb. of pure surfactant.
4. Recent developments in surfactants solutions for EOR have effectively reduced the required surfactant concentration which lowering the chemical costs required.
5. Recently, new and effective surfactants are derived from plant resources such as sunflower oil, soy and corn oil. It is non-toxic, non-hazardous, and readily biodegradable.

The disadvantages of surfactant flooding could be listed as following:

1. Complex and expensive system.
2. Possibility of chromatographic separation of chemicals.
3. High adsorption of surfactant.
4. Losing its effectiveness at higher pressure, temperature, and salinity.

2.2.3 Field projects of the surfactant flooding

Many technically successful pilots have been done in addition to several small commercial projects have been completed and several more are in progress. Relatively, homogeneous reservoir formation is preferred. The presence of high amounts of clays, gypsum, or anhydrite is undesirable. For commercially available surfactants, formation-water chlorides should be less than 20,000 ppm and divalent ions (Ca^{++} and Mg^{++}) should be less than 500 ppm. The problems encountered with some of the old pilots are well understood and have been solved and the new generation surfactants will tolerate high salinity and high hardness so there is no practical limit for high salinity reservoirs [14].

2.3 Surfactant-polymer (SP) flooding

Surfactant-polymer flooding process is injecting a chemical slug that contains water, surfactant, electrolyte (salt), usually a co-surfactant (alcohol), followed by polymer-thickened water. In this process a surfactant is added to the polymer solution that has the affinity for both water and oil. The use of the micellar solution is to reduce the interfacial tension of the water-oil system in the reservoir in order to displace the residual oil [15]. SP flooding method was patented for Marathon oil co. by Gogarty and Tosch known as Mara-flood. The injection profile of the method consists of injecting a pre-flush (to achieve the desired salinity environment), followed by micellar slug (surfactant, co-surfactant, electrolyte), and followed by polymer solution along with drive water.

The micellar solution composition that ensures a gradual transition from the displacement water to the displaced oil without interface is as following [13]:

- Surfactant 10–15%.
- Water 20–60%.
- Oil 25–70%.
- Co-surfactant 3–4%.

Usually, the co-surfactant is alcohol which enhances the possibility for the micellar solution to include oil or water. This surfactant-polymer flooding reduces the oil-water IFT through the surfactant portion and reduces the mobility ratio through presence of polymer.

2.3.1 Mechanism

The micellar solution is prepared using inorganic salts (water-soluble electrolytes) in order to gain better viscosity control of the solution. A polymer slug is used to drive the micellar solution slug in order to get a mobility control. The injection process is shown in **Figure 4**. The technique establishes low oil-water IFT and controls the mobility ratio which forming a considerable oil bank to be produced [11].

2.3.2 Surfactant-polymer flooding advantages

The SP flooding advantages are listed below:

- Interfacial tension reduction (improves displacement sweep efficiency).
- Mobility control (improves volumetric sweep efficiency).
- Reduce adsorption of expensive surfactants.

The disadvantages could be as follows:

- Complex and expensive system.
- Possibility of chromatographic separation of chemicals.

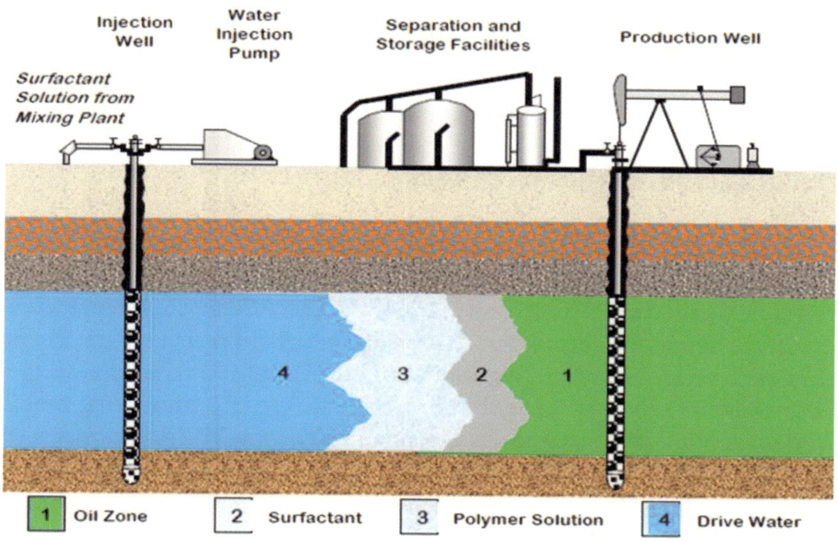

Figure 4.
Surfactant-polymer injection process.

- High adsorption of surfactant.
- Interactions between surfactant and polymer.
- Degradation of chemicals at high temperature.

2.3.3 Field project of the surfactant-polymer flooding

Since 1990, polymer flood and SP flood have been applied in a few field pilots and expanded field tests.

2.4 Alkaline flooding

Alkaline flooding is one of the EOR methods in which alkaline agents are injected into the reservoir to produce in situ surfactants, so the alkaline flooding will eventually have the same effect of the surfactant flooding.

2.4.1 Mechanism

In the Alkaline flooding process, the alkaline agents such as sodium hydroxide solution is injecting into the reservoirs which react with the naturally occurring organic acids in the oil in order to produce surfactants or soaps at the oil-water interface. However, the alkaline agents are less expensive than the surfactant agents, the expected incremental oil recovery by alkaline flooding has not been confirmed by field results and still remains possibility as the process is mainly dependent on the mineral composition of the reservoir rock and its oil [11].

2.4.2 Alkaline flooding advantages

This EOR method has the same advantages of the surfactant flooding in addition to that its main advantage over the surfactant is the cost of the alkaline agents are cheap compared to the surfactant agents [12].

2.4.3 Field project of the surfactant-polymer flooding

There were several pilot tests worldwide such as in Russian Tpexozephoe Field, Hungarian H Field, Whittier Field in California, and North Gujarat Oil Field, India.

2.5 Alkaline-surfactant-polymer (ASP) flooding

Individual chemical flooding processes, alkaline flooding, surfactant flooding and polymer flooding, can be combined differently. The three-component combination, alkaline surfactant-polymer (ASP). The ASP method represents a cost-effective chemical EOR method that yielding high oil recovery (mostly for sandstone reservoirs). ASP flooding is utilizing the benefits of three flooding methods, where oil recovery was enhanced, by reducing IFT, improving mobility ratio, and improving microscopic displacement efficiency. The ASP projects in China shows that the incremental oil recovery over water-flooding is 18.9% on the average [14].

2.5.1 Mechanism

Alkaline injection reduces surfactant adsorption and the combination of soap and synthetic surfactant results in low interfacial tension (IFT) in a wider range of salinity. Soap and surfactant make emulsions stable through reduced IFT which improve the sweep efficiency. There is a competition of adsorption sites between polymer and surfactant. Therefore, addition of polymer reduces surfactant adsorption, or vice versa and improves the sweep efficiency of ASP solution [5].

2.5.2 Alkaline surfactant-polymer flooding advantages

Several advantages can be summarized as follows:

- Alkali is inexpensive, so it is cost reduction factor.
- Alkali reacts with acid in oil to form soap.
- Provide lower IFT in a wide salinity range.
- Soaps and surfactants produce emulsions that improve the sweep efficiency.
- Polymer and alkaline are reducing the surfactant adsorption.
- The polymer addition improves the sweep efficiency of the ASP solution.
- Carbonate formations are usually positively charged at neutral pH, which favors adsorption of anionic surfactants. However, when (Na_2CO_3) is present, carbonate surfaces (calcite, dolomite) become negatively charged and adsorption decreases several fold.
- High pH also improves micro-emulsion phase behavior.

The limitations and challenges for ASP flooding are:

- Severe scaling in the injection lines with strong emulsification of the produced fluid.

- Polymers are less effective under high water salinity conditions, as the high salt waters degrade the viscosity of polymers.

- Mobility control is critical.

- Laboratory tests must be done with crude and reservoir rock under reservoir conditions and are essential for each reservoir condition.

2.5.3 Field projects of the alkaline surfactant-polymer flooding

There are large field trials that already implemented worldwide showing encouraging results. The following table (**Table 3**) shows a summary for the ASP projects or underway since 1980 including the start-up date, oil gravity, Oil viscosity, implementation phase as secondary or tertiary, oil recovered in % of OOIP, and the chemical cost in USD/bbl. In **Figure 3**, the production results after applying the ASP flooding at the end of the water-flooding phase [13].

Surfactant Enhanced Water Floods

Alkaline Surfactant Polymer Projects Completed or Underway Since About 1980

Field	Owner	Techical	Region	Start	API	Oil Viscosity - cp	Type	Pore Volume Chemical	Oil Recovered %OOIP	Chemicals US Cost/bbl	
Adena	Babcock & Brown	Surtek	Colorado	2001	43	0.42	Tertiary	In progress		$2.45	Na2CO3
Cambridge	Barrett	Surtek	Wyoming	1993	20	25	Secondary	60.4%	28.07%	$2.42	Na2CO3
Cressford	Dome	Surtek	Alberta	1987			Secondary			$2.25	Alkali and Polymer Only
Daquing BS	Sinopec		China	1996	36	3	Tertiary	82.1%	23.00%	$7.88	NaOH - Biosurfactant
Daquing NW	Sinopec	Surtek	China	1995	36	3	Tertiary	65.0%	20.00%	$7.80	NaOH
Daquing PO	Sinopec	Surtek	China	1994	26	11.5	Tertiary	42.0%	22.00%	$5.51	Na2CO3
Daquing XV	Sinopec		China		36	3	Tertiary	48.0%	17.00%	$9.26	NaOH
Daquing XF	Sinopec		China	1996	36	3	Tertiary	55.0%	25.00%	$7.14	NaOH
Daquing Foam	Sinopec		China	1997	NA	NA	Tertiary	54.8%	22.32%	$8.01	ASPFoam Flood following WAG
Daquing Scale Up	Sinopec		China	??	Reported to be Shut In Due to QC Problems with Surfactant						
David		Surtek	Alberta	1985	23		Tertiary			$0.80	
Driscoll Creek	TRUE	Surtek	Wyoming	1998	Acrylamid converted to acrylate - water cut lowered						
Enigma	Citation	Surtek	Wyoming	2001	24	43	Secondary	In progress		$2.49	Na2CO3
Etzikom	Renaissance/Husky	Surtek	Alberta	Current	In progress - Information not released						
Gudong	CNPC	Shenli	China	1992	17.4	41.3	Tertiary	55.0%	26.51%	$3.92	
Isenhaur	Enron	Tiorco	Wyoming	1980	43.1	2.8	Secondary	57.7%	11.58%	$0.83	Alkali and Polymer Only
Karmay	CNPC	UT/NIPER	China	1996	30.3	52.6	Tertiary	60.0%	24.00%	$4.35	
Lagomar	PDVSA	Surtek	Venezuela	2000	24.8	14.7	Tertiary	45.0%	20.11%	$4.80	Single Well Test
Mellot Ranch	West	Surtek	Wyoming	2000	22	23	Tertiary	In progress		$2.51	NaOH
Minas I	Chevron	Chevron	Indonesia	1999	Micellar Polymer Failed when salinity of slug decreased						
Minas II	Texaco	Texaco	Indonesia	Current	Lignin II Surfactant - In progress - Information not released						
Sho Vel Tum	LeNorman	DOE	Oklahoma	1998	26.4	41.3	Tertiary	60.0%	16.22%	$6.40	Low Acid Number - Viscous
Bevery Hills	Stocker	Tiorco	California	Surfactant Injectivity Test							
Tanner	Citation	Surtek	Wyoming	2000	21	11	Secondary	In progress		$2.82	NaOH
West Kiehl	Barrett	Surtek	Wyoming	1987	24	17	Secondary	26.5%	20.68%	$2.13	
West Moorcroft	KSL	Tiorco	Wyoming	1991	22.3	20	Secondary	20.0%	15.00%	$1.46	Alkali and Polymer Only
White Castle	Shell	Shell	Louisiana	1987	29	2.8	Tertiary	26.9%	10.10%	$8.18	*No Polymer*

Table 3.
Field cases of ASP EOR.

Author details

Ahmed Ragab[1] and Eman M. Mansour[2]*

1 Reservoir Department, Agiba Petroleum Company, Cairo, Egypt

2 Production Department, Egyptian Petroleum Research Institute, Cairo, Egypt

*Address all correspondence to: emanmansour84@yahoo.com

References

[1] Mansour E, Al-Sabagh A, Desouky S, Zawawy F, Ramzi M. A laboratory investigation of carbon dioxide-enhanced oil recovery by focusing on CO_2-oil physical properties. Egyptian Journal of Petroleum. 2019;**28**(1):21-26

[2] Tunio SQ, Tunio AH, Ghirano NA, El Adawy ZM. Comparison of different enhanced oil recovery techniques for better oil productivity. International Journal of Applied Science and Technology. 2011;**1**(5)

[3] Manrique EJ et al. EOR: Current status and opportunities. In: SPE Improved Oil Recovery Symposium. Tusla, USA: Society of Petroleum Engineers (SPE); 2010

[4] Mansour E, Al-Sabagh A, Desouky S, Zawawy F, Ramzi M. A new estimating method of minimum miscibility pressure as a key parameter in designing CO_2 gas injection process. Egyptian Journal of Petroleum. 2017

[5] Raffa P, Broekhuis AA, Picchioni F. Polymeric surfactants for enhanced oil recovery: A review. Journal of Petroleum Science and Engineering. 2016;**145**:723-733

[6] Abidin A, Puspasari T, Nugroho W. Polymers for enhanced oil recovery technology. Procedia Chemistry. 2012;**4**:11-16

[7] Cankara I. Polymer/Oil Relative Permeabilities in Carbonate Reservoirs [M. S. thesis]. Ankara, Turkey: Middle East Technical University; 2005

[8] Sun C, Hou J, Pan G, Xia Z. Optimized polymer enhanced foam flooding for ordinary heavy oil reservoir after cross-linked polymer flooding. Journal of Petroleum Exploration and Production Technology. 2016;**6**(4):777-785

[9] Mansour E, Al-Sabagh A, Desouky S, Zawawy F, Ramzi M. Experimental approach of minimum miscibility pressure for CO_2 miscible flooding: Application to Egyptian oil fields. International Journal of New Technology and Research. 2016;**2**(5)

[10] Sheng J. Modern Chemical Enhanced Oil Recovery: Theory and Practice. Oxford, United Kingdom: Gulf Professional Publishing; 2010

[11] Cockcroft P, Anli J, Duignan J. EOR Potential of Indonesian Reservoirs; Annual Convention Proceedings, IPA 1988;**03**:21-88

[12] Sheng J. Enhanced Oil Recovery Field Case Studies. Oxford, United Kingdom: Gulf Professional Publishing; 2013

[13] Sheng JJ. Status of surfactant EOR technology. Petroleum. 2015;**1**(2):97-105

[14] Taber JJ, Martin F, Seright R. EOR screening criteria revisited—Part 1: Introduction to screening criteria and enhanced recovery field projects. SPE Reservoir Engineering. 1997;**12**(03):189-198

[15] Karović-Maričić V, Leković B, Danilović D. Factors influencing successful implementation of enhanced oil recovery projects. Podzemni radovi. 2014;**22**(25):41-50

Section 2

Ocean Waves Studies

Chapter 5

High-Resolution Numerical Simulation of Surface Wave Development under the Action of Wind

Dmitry Chalikov

Abstract

The paper describes the numerical experiments with a three-dimensional phase-resolving model based on the initial potential equation of motion with free surface at deep water in the periodic domain written in the surface-following nonstationary curvilinear nonorthogonal coordinate system. The numerical scheme is based on Fourier-transform method. The vertical velocity on surface is calculated by solving the three-dimensional Poisson equation for the velocity potential. The velocity potential is represented as a sum of linear and nonlinear components. The linear component is described by Laplace equation. The nonlinear component is calculated by solution of the three-dimensional Poisson equation with the iterated right-hand side. The model includes some algorithms for calculation of the energy input from wind as well as for calculation of breaking and high-frequency dissipation. Initially, the conditions are assigned as a set of small waves corresponding to JONSWAP spectrum at high wave number. In response to waves' growth, the spectrum shifts to lower wave numbers. The evolution of spectrum is generally in an agreement with the observed data. The wave spectrum and the spectra of different rates of energy transformation as well as the statistical characteristics of wave field for different stages of development are described.

Keywords: numerical simulation, wind waves, waves' development, wave spectrum, Fourier-transform method, wind input, waves' dissipation, wave statistics

1. Introduction

The development of waves under the action of wind is a process that is difficult to simulate since surface waves are very conservative and their energy changes over hundreds and thousands of periods. This is why the most popular method is spectral modeling based on the averaged over phase equations for spectral energy. In this approach, the waves as physical objects are actually absent since the evolution of spectral distribution of the wave energy is simulated. The description of input and dissipation in this approach is not directly connected with the formulation of the problem, but it is rather adopted from other branches of the wave theory where waves are the objects of investigation. However, the spectral approach was found to be the only method capable to describe the space and time evolution of wave field in the ocean. The phase resolving models (or 'direct' models) designed for reproducing waves themselves cannot compete with spectral models since such models typically can reproduce the evolution of just several thousands of large waves. Nevertheless, the direct wave modeling plays an ever-increasing role in the geophysical fluid dynamics because it gives the possibility to investigate the processes that cannot be reproduced by spectral models.

The spectral model assumes that wave field consists of a superposition of linear waves with random phases and arbitrary angle distribution. Being converted to a physical wave field, it looks unreal because real waves usually have prolonged smooth troughs and sharp peaks. Such shape suggests that the waves are similar to Stokes waves. For any given wave spectrum, the wave field can be represented as a superposition of linear Fourier modes with random phases [1]. It can be represented also as a superposition of Stokes modes. The calculations of statistical characteristics for both wave fields show that they are nearly identical. However, such conclusion could be made with no calculations because typical steepness of sea waves at reasonable spectral resolution is of the order of 0.01–0.05, so all the amplitudes of Stokes modes starting from the second one are small. It follows that the specific shape of sea waves is a dynamic property.

The breaking is usually initiated in the vicinity of wave peaks, so the breaking parameterization describes the processes, which are, in principle, impossible in a linear wave field. The breaking is concentrated in the separated narrow intervals; an instantaneous spectrum is discrete and is shifted to high frequencies. The spectral description of the isolated extreme waves is also impossible. The Fourier transform of a large-scale wave field including separated large waves does not provide any indications of their appearance.

The input energy to waves is based on the assumption that each mode induces a pressure mode with a certain amplitude and phase. In reality, the pressure field is quite complicated and not directly connected with the surface elevation because of the systematic separation of air flow behind wave crests (which was shown experimentally in Refs. [2, 3] with the coupled wind-wave model). There are many other complications in the coupled wind-wave dynamics [4] including the processes connected with sprays, bubbles, foam, and structure of the high-frequency wave spectrum.

Most (but not all) of the processes mentioned above can be investigated using the numerical modeling that is a perfect instrument for development of parameterization of physical processes for spectral wave models.

The phase-resolving models can completely replace the spectral models for direct simulation of wave regimes of small water basins, for example, port harbors (see Refs. [5, 6]). Other approaches of direct modeling are discussed in Refs. [7, 8].

Over the past decades, a big volume of papers devoted to the numerical methods developed for investigation of wave processes has been published. The most advanced among them are Finite Difference Method [5, 6], Finite Volume Method [9], Finite Element Method [10, 11], Boundary (Integral) Element Method [12], Spectral HOS Methods [13–17], the Smoothed Particle Hydrodynamics Method [18], Large Eddy Simulation Method (LES) [19, 20], Moving Particle Semi-Implicit Method [21], Constrained Interpolation Profile Method [22], and Method of Fundamental Solutions [23]. Most of the models were designed for engineering application handling such processes as overturning waves, broken waves, waves generated by landslides, freak waves, solitary waves, tsunamis, violent sloshing waves, interaction of extreme waves with beaches, as well as interaction of steep waves with fixed and different floating structures. The wave models designed for engineering applications seem to be more advanced than the models for pure geophysical research. However, as a rule, the engineering models pay little attention to description of physical processes that are responsible for a long-term evolution of wave spectrum. A more detailed review of direct numerical models is given in Ref. [8].

Until recently, the direct modeling was used for reproduction of a quasi-stationary wave regime when wave spectrum does not change significantly. An example of direct numerical modeling of surface wave evolution is given in Ref. [24] where the development of wave field was calculated by using of a two-dimensional model based on full potential equations written in the conformal coordinates. A model included the algorithms for parameterization of the input and dissipation of energy (a description of similar algorithms is given below). The model successfully reproduced the evolution of wave spectrum under the action of wind. That model was a prototype of 3-D model, because being very fast it was convenient for development of the physical process parameterization. However, the strictly one-dimensional (unidirected) waves are not quite realistic since the unidirected waves in the presence of the small-amplitude perturbations relatively quickly turn into the two-dimensional wave field [25]. Hence, the full problem of wave evolution should be formulated on the basis of three-dimensional equations. Such 3-D calculations were done by Chalikov [26]. The model included parameterization of the main physical processes: input energy, different types of dissipation, and transformation of spectrum due to the nonlinear interaction. The last process does not require any parameterization because the nonlinearity is described with equations. The model used a relatively poor resolution (1024 $\times$ 512 nodes in x and y directions). However, the calculations reproduced the evolution of wave spectrum and the spectra of the main physical processes such as input and dissipation of energy and nonlinear interactions. As long as we know, it was the first attempt to reproduce the development of waves based on full three-dimensional equations with a direct solution of 3-D equation for the velocity potential. The current paper is devoted to development of the method, including the tuning and modifications of the algorithm, the increase of resolution, and the integration for longer periods. The most important but pure technical modifications were introduced in the numerical scheme for Poisson equation for the velocity potential. The algorithms for calculation of the input and dissipation remain nearly the same, but the numerical parameters in those schemes were changed to achieve a better agreement with the rate of spectrum evolution given by JONSWAP approximation.

2. Equations

The *nonstationary surface-following nonorthogonal* coordinate system is used as follows:

$$\xi = x, \quad \vartheta = y, \quad \zeta = z - \eta(\xi, \vartheta, \tau), \quad \tau = t, \tag{1}$$

where $\eta(x, y, t) = \eta(\xi, \vartheta, \tau)$ is a moving periodic wave surface given by the Fourier series

$$\eta(\xi, \vartheta, \tau) = \sum_{-M_x < k < M_x} \sum_{-M_y < l < M_y} h_{k,l}(\tau)\Theta_{k,l}, \tag{2}$$

where k and l are the components of the wave number vector k; $h_{k,l}(\tau)$ are Fourier amplitudes for elevations $\eta(\xi, \vartheta, \tau)$; M_x and M_y are the numbers of modes in the directions ξ and ϑ, respectively, while $\Theta_{k,l}$ are the Fourier expansion basis functions represented as the matrix:

$$\Theta_{kl} = \begin{cases} \cos(k\xi + l\vartheta) & -M_x \le k \le M_x, \quad -M_y < l < 0 \\ \cos(k\xi) & -M_x \le k \le 0, \ l = 0 \\ \sin(k\xi) & 0 \le k \le M_y, \ l = 0 \\ \sin(k\xi + l\vartheta) & -M_x \le k \le M_x, \ 0 < l \le M_y \end{cases} \tag{3}$$

The 3-D equations of potential waves in the system of coordinates (1) at $\zeta \le 0$ take the following form:

$$\eta_\tau = -\eta_\xi \varphi_\xi - \eta_\vartheta \varphi_\vartheta + \left(1 + \eta_\xi^2 + \eta_\vartheta^2\right)\Phi_\zeta \tag{4}$$

$$\varphi_\tau = -\frac{1}{2}\left(\varphi_\xi^2 + \varphi_\vartheta^2 - \left(1 + \eta_\xi^2 + \eta_\vartheta^2\right)\Phi_\zeta^2\right) - \eta - p, \tag{5}$$

$$\Phi_{\xi\xi} + \Phi_{\vartheta\vartheta} + \Phi_{\zeta\zeta} = \Upsilon(\Phi), \tag{6}$$

where Υ is the operator:

$$\Upsilon() = 2\eta_\xi()_{\xi\zeta} + 2\eta_\vartheta()_{\vartheta\zeta} + (\eta_{\xi\xi} + \eta_{\vartheta\vartheta})()_\zeta - \left(\eta_\xi^2 + \eta_\vartheta^2\right)()_{\zeta\zeta}, \tag{7}$$

capital fonts Φ are used for the domain $\zeta < 0$, while the lower case φ refers to $\zeta = 0$. The term p in Eq. (5) describes the pressure on the surface $\zeta = 0$.

It is suggested by Chalikov et al. [7] that it is convenient to represent the velocity potential φ as a sum of the two components: a linear component $\overline{\Phi}$, $(\overline{\varphi} = \overline{\Phi}(\xi, \vartheta, 0))$ and an arbitrary nonlinear component $\tilde{\Phi}$, $(\tilde{\varphi} = \tilde{\Phi}(\xi, \vartheta, 0))$:

$$\varphi = \overline{\varphi} + \tilde{\varphi}, \quad \Phi = \overline{\Phi} + \tilde{\Phi}. \tag{8}$$

The linear component $\overline{\Phi}$ satisfies Laplace equation:

$$\overline{\Phi}_{\xi\xi} + \overline{\Phi}_{\vartheta\vartheta} + \overline{\Phi}_{\zeta\zeta} = 0, \tag{9}$$

with a known solution:

$$\overline{\Phi}(\xi, \vartheta, \zeta, \tau) = \sum_{k,l} \overline{\varphi}_{k,l}(\tau) \exp(|k|\zeta)\Theta_{k,l}, \tag{10}$$

where $|k| = \left(k^2 + l^2\right)^{1/2}$, $\overline{\varphi}_{k,l}$ are the Fourier coefficients of the surface linear potential $\overline{\varphi}$ at $\zeta = 0$). The solution satisfies the following boundary conditions:

$$\begin{aligned} \varsigma = 0: &\quad \overline{\Phi} = \overline{\varphi} \\ \varsigma \to -\infty: &\quad \tilde{\Phi}_\zeta \to 0 \end{aligned} \tag{11}$$

The nonlinear component satisfies the equation:

$$\tilde{\Phi}_{\xi\xi} + \tilde{\Phi}_{\vartheta\vartheta} + \tilde{\Phi}_{\zeta\zeta} = \Upsilon\left(\tilde{\Phi}\right) + \Upsilon\left(\overline{\Phi}\right), \tag{12}$$

Eq. (12) is solved with the boundary conditions:

$$\begin{aligned} \varsigma = 0: &\quad \tilde{\Phi} = 0 \\ \varsigma \to -\infty: &\quad \tilde{\Phi}_\zeta \to 0 \end{aligned} \tag{13}$$

The presentation (8) is not used for solution of the evolutionary Eqs. (4) and (5) because it does not provide any improvements of accuracy and speed.

Eqs. (4)–(6) are written in a nondimensional form by using the following scales: length L, where $2\pi L$ is a (dimensional) period in the horizontal direction; time $L^{1/2}g^{-1/2}$; and velocity potential $L^{3/2}g^{1/2}$ (g is the acceleration of gravity). The pressure is normalized by the water density, so that the pressure scale is Lg. Eqs. (4)–(6) are self-similar to the transformation with respect to L. The dimensional size of the domain is $2\pi L$, so the scaled size is 2π. All of the results presented in this paper are nondimensional. Note that the number of the Fourier modes can be different in the x and y directions. In this case, it is assumed that the two-length scales L_x and L_y are used. The nondimensional length of the domain in the y-direction remains equal to 2π, and a factor $r = L_x/L_y$ is introduced into the definition of a differential operator in the Fourier space.

The derivatives of a linear component $\overline{\Phi}$ in (7) are calculated analytically. The scheme combines a 2-D Fourier transform method in the 'horizontal surfaces' and a second-order finite-difference approximation on the stretched staggered grid defined by the relation $\Delta\zeta_{j+1} = \chi\Delta\zeta_j$ ($\Delta\zeta$ is a vertical step, while $j = 1$ at the surface). The stretched grid provides an increase in accuracy of approximation for the exponentially decaying modes. The values of the stretching coefficient χ lie for different settings within the interval 1.01–1.20; in the current work, the value$\gamma = 1.2$ was used at the number of levels $L_w = 10$. Such poor resolution was possible to use because of the separation of the potential into a large linear and a small nonlinear part, so Eq. (12) was used only for calculation of a small correction for the potential. A high value of the stretching coefficient provided high resolution in the vicinity of surface for accurate calculations of the surface vertical derivative for the potential. The finite-difference second-order approximation of vertical operators in Eq. (12) on a nonuniform vertical grid is quite straightforward (see Ref. [8]). Eq. (12) is solved as Poisson equations with the iterations over the right-hand side by TDMA method [27]. At each time step, the iterations start with a right-hand side calculated at the previous time step. A relative accuracy of the solution in terms of the vertical derivative of the potential on the surface was equal to 10^{-6}. The typical number of iterations was 2–5.

The accuracy of the adiabatic version of equations was validated by reproducing a moving Stokes wave with the steepness $AK = 0.40$ (A is a half of trough-to-crest wave height; $K = 1$ is the wave number of the first mode). An algorithm for calculation of Stokes wave with the prescribed accuracy was suggested by Chalikov and Sheinin [28]. The scheme based on the conformal coordinates is very effective: the calculations were carried out in 100 ms at notebook (2.10 GHz). The dependence of Stokes wave spectra on the wave number is shown in **Figure 1**. In fact, about 2000 curves obtained in the course of calculations, with the interval $\Delta t = 1$, were plotted. Due to improvement of the numerical scheme, the accuracy of reproduction of Stokes wave is considerably higher than for the scheme used in Refs. [7, 8].

As seen, up to $S \sim 10^{-12}$ ($k = 22$), the spectra of Stokes waves remain with high accuracy the same as it was assigned in initial conditions. At higher frequencies, the random disturbances appear. Note that this validation is not trivial: even small inaccuracies in the numerical scheme cause a fast distortion of the spectra, like in the bottom part of **Figure 1**. We consider these results as a serious evidence of high accuracy of the adiabatic version of the model. The previous version of 3-D model [26] allowed carrying out a long simulation of Stokes wave not steeper than $AK = 0.30$. The right-hand sides of Eqs. (4) and (5) were calculated with a use of Fourier transform method: the nonlinear terms were calculated at the extended grid with size $(4M_x \times 4M_y)$, and then by the inverse Fourier transform, they were returned

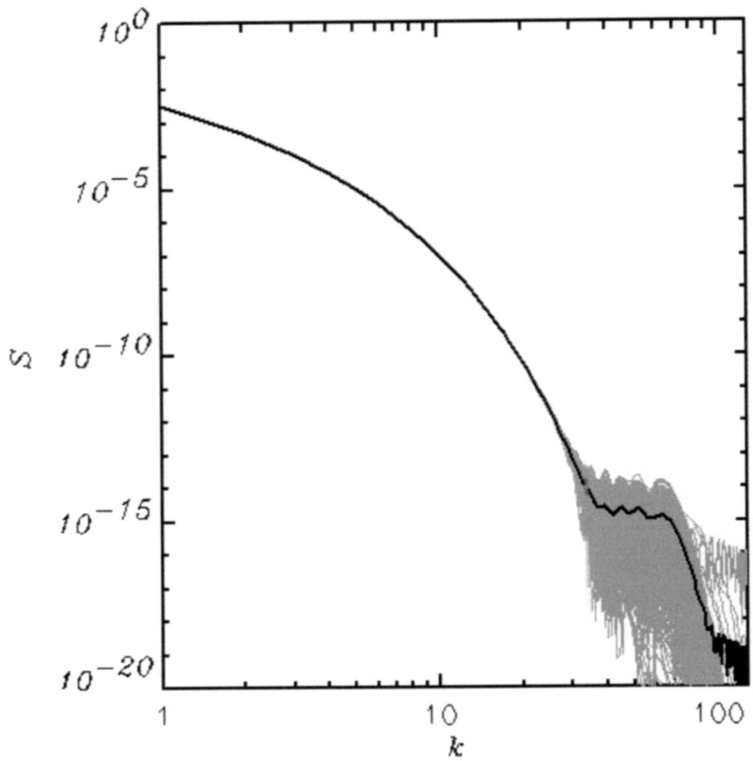

Figure 1.
The 2000 spectra of Stokes wave with steepness AK = 0.40 as a function of wave number k superimposed on each other. The gray area in the right left corner is a computational noise.

to the Fourier grid. The fourth-order Runge-Kutta scheme was used for integration in time. The equation for the potential was solved at each of the four substeps of time step.

The simulations described by Chalikov [26] were a first attempt to reproduce the development of wave field assigned in the initial conditions as a group of small waves at high wave number under the action of strong wind. The initial elevation was generated as a superposition of linear waves corresponding to JONSWAP spectrum [29] with random phases. The initial Fourier amplitudes for the surface potential were calculated by the formulas of the linear wave theory. The details of the initial conditions are of no importance because the initial energy level is quite low. The wave peak was placed to the wave number equal to 100. The wind velocity was assigned equal to$4c_{100}$, where c_{100} is a phase velocity of the 100th mode. A detailed description of the scheme and its validation is given in Refs. [7, 8].

The simulation described in the current paper was performed with a doubled resolution in both directions, with the improved numerical scheme for Poisson equations and modified parameters in the scheme for calculations of energy transitions.

3. Energy input

The detailed description of the algorithm for calculation of energy input is given in Ref. [26]. The energy and momentum are transferred from air to water by the surface pressure field and tangent stress. According to the most reasonable theory

[30], the Fourier components of surface pressure p are connected with those of the surface elevation through the following expression:

$$p_{k,l} + ip_{-k,-l} = \frac{\rho_a}{\rho_w}(\beta_{k,l} + i\beta_{-k,-l})(h_{k,l} + ih_{-k,-l}), \tag{14}$$

where $h_{k,l}$, $h_{-k,-l}$, $\beta_{k,l}$, $\beta_{-k,-l}$ are real and imaginary parts of elevation η, and the so-called β-function, ρ_a/ρ_w, is the ratio of air and water densities. Both β coefficients are the functions of the nondimensional frequency

$$\Omega = \omega U/g, \tag{15}$$

that characterizes the ratio of wind velocity to phase velocity of c_k:

$$\Omega = U/c_k \tag{16}$$

Since the supplying of wave with the energy and momentum occurs in a layer whose height is proportional to the wave length, it is reasonable to suggest that the reference height for the wind velocity should be different for a different virtual wave length (distance $\lambda_k/\cos\theta_i$ between the wave peaks in wind direction; the index i denotes a direction of mode). The wind velocity can be found by interpolation or extrapolation to the level:

$$z_{i,k} = 0.5\lambda_k/\cos\theta_i \tag{17}$$

The definition of Ω_κ should take into account the angle θ_i between the vector U and the direction of wave mode. Finally, the virtual nondimensional frequency takes the form:

$$\Omega_{i,k} = \omega_k \cos\theta_i U(z_k)/g = \cos\theta_i U(z_k)/c_k \tag{18}$$

where $c_k = g/\omega_k$ is the phase velocity of kth mode.

For experimental derivation of the shape of β-function, it is necessary to simultaneously measure the wave surface elevation and nonstatic pressure on the surface [31–35]. The data obtained in this way allowed constructing an imaginary part of β-function used in some versions of the wave forecasting models [36]. The data on experimental β-function are compared in Ref. [4]. The values of β within the interval $(0 < \Omega < 10)$ differ by decimal orders. Hence, the question arises: in what way, using such a different input, the spectral models provide a reasonable agreement with the observations. The answer is very simple: the researchers have the possibility to modify the parameterization of dissipation. Despite the hundreds of papers, the knowledge on dissipation is even poorer than the knowledge on the energy input. Finally, only the sum of those source terms regulates the growth of total wave energy. Such situation is far from being perfect since the energy input and dissipation have totally different spectral properties.

The second way of the β-function evaluation is based on the results of numerical investigations of the statistical structure of the boundary layer above waves with the use of Reynolds equations and an appropriate closure scheme. In general, this method works so well that many problems in the technical fluid mechanics are often solved not experimentally but by using the numerical models [37, 38]. This method was being developed beginning from Refs. [39, 40] and followed by Refs. [41–43]. The results were implemented in the WAVEWATCH model, i.e., the third-generation wave forecast model [44], and thoroughly validated against the experimental data in the course of developing WAVEWATCH-III [45]. Most of the schemes for the calculations of β-function consider a relatively narrow interval of the nondimensional frequencies Ω. In the current work, the range of frequencies covers the interval $(0 < \Omega < 10)$, and occasionally, the values of $\Omega > 10$ can appear.

The most reliable data on β-function are concentrated in the interval $-10 < \Omega < 10$ (the negative values of Ω correspond to the wave modes running against wind). In the current calculations, the modes running against wind are absent. The function β can be approximated by the formulas:

$$\beta_i = \begin{cases} \beta_0 + a_0(\Omega - \Omega_0) + a_1(\Omega - \Omega_0)^2 & \Omega > \Omega_0 \\ \beta_0 - a_0(\Omega - \Omega_0) + a_1(\Omega - \Omega_0)^2 & \Omega < \Omega_0 \end{cases}, \tag{19}$$

$$\beta_r = \begin{cases} \beta_1 + a_3(\Omega - \Omega_2) & \Omega < \Omega_2 \\ a_2(\Omega - \Omega_1)^2 & \Omega_2 < \Omega < \Omega_3 \\ \beta_1 - a_3(\Omega - \Omega_3) & \Omega > \Omega_3 \end{cases}, \tag{20}$$

where $\Omega_0 = 0.355$, $\Omega_1 = 1.20$, $\Omega_2 = -18.8$, $\Omega_3 = 21.2$, $a_0 = 0.0228$, $a_1 = 0.0948$, $a_2 = -0.372$, $a_3 = 14.8$, $\beta_0 = -0.02$, $\beta_1 = -148.0$.

The wind velocity remains constant throughout the integration. The values of Ω for other wave numbers are calculated by assuming that the wind profile is logarithmic.

Note that the formulation of wind and waves interaction can be significantly improved by coupling the wave model with the 1-D Wave Boundary Layer model [4]. The next step can be the coupling of wave model with the 3-D model of WBL based on the closure schemes or LES model (see Ref. [46]).

4. Energy dissipation

The current version of the model includes three types of dissipation (see details in Ref. [26]).

1. The energy can decrease due to the errors of approximations in space and time that depend on the number of Fourier modes, number of knots in the physical space, the vertical grid used for approximation of Poisson equation (6), and the criterion for accuracy of its solution. All of those errors that produce the 'numerical dissipation' can be referred to the adiabatic part of the models (4)–(6) at $p = 0$. The rate of this dissipation can be reduced by the use of a better resolution and a higher accuracy of approximation, but this way leads to deceleration of the calculations with the model already running for a very long time.

   Opposite to the numerical dissipation, there exists another type of energy loss that has rather a physical nature. The nonlinear interaction of different modes forms a flux of energy directed outside of the computational domain. We call it the 'nonlinear dissipation.' The numerical and nonlinear dissipation can hardly be considered separately. The estimation of rate of the numerical/nonlinear dissipation can be easily done by the comparison of full energy before and after the time step for the adiabatic part of the model (see Section 4 in Ref. [26]). In the current calculations, the loss of energy for one time step was about 10^{-4}%, which is by 2–3 orders less that the rate of energy change due to input energy. Since we prefer to consider the process described by Eqs. (4)–(6) as adiabatic one, at each time step we restore the energy lost by both the numerical and nonlinear dissipation.

2. A long-term integration of full fluid mechanics equations always shows the spreading of spectrum to both high and low frequencies (wave numbers). The nonlinear flux of energy directed to the small wave numbers produces downshifting of spectrum, while an opposite flux forms a shape of the spectral tail. The second process that we call the 'tail dissipation' can produce accumulation of energy near the 'cut' wave number. The growth of amplitudes at high wave numbers is followed by growth of the local steepness and development of the numerical instability. To support the stability, additional terms are included into the right-hand sides of Eqs. (4) and (5):

$$\frac{\partial \eta_{k,l}}{\partial \tau} = E_{k,l} - \mu_{k,l}\eta_{k,l}, \tag{21}$$

$$\frac{\partial \varphi_{k,l}}{\partial \tau} = F_{k,l} - \mu_{k,l}\varphi_{k,l} \tag{22}$$

(where $E_{k,l}$ and $F_{k,l}$ are the Fourier amplitudes of the right-hand sides of Eqs. (4) and (5); the value of $\mu_{k,l}$ is equal to zero inside the ellipse with semi-axes $d_m M_x$ and $d_m M_y$; then, it grows quadratically with $|k|$ up to the value c_m and is equal to c_m outside of the outer ellipse (see details in Ref. [26]). This method of filtration that we call the 'tail dissipation' was developed and validated with the conformal model [28]. The sensitivity of the results to the parameters in Refs. (21) and (23) is not large. The aim of the algorithm is to support smoothness and monotonicity of the wave spectrum within the high wave number range.

3. The main process of wave dissipation is the 'breaking dissipation.' This process is taken into account in all the spectral wave forecasting models similar to WAVEWATCH (see Refs. [44, 47]). Since there are no waves in the spectral models, no local criteria of wave breaking can be formulated. This is why the breaking dissipation is represented in the spectral models in a distorted form. The real breaking occurs in the relatively narrow areas of the physical space; however, the spectral image of such breaking is stretched over the entire wave spectrum, while in reality, the breaking decreases height and energy of separate waves. This contradiction occurs because the waves in the spectral models are assumed to be linear. In fact, a nonlinear sharp wave breaks in the physical space. Such wave is often composed of several local modes. It is clear that the state-of-art wave models should account for the threshold behavior of a breaking wave, that is, waves will not break unless their steepness exceeds the threshold [48–50].

The instability of the interface leading to breaking is an important though poorly developed problem of fluid mechanics. In general, this essentially nonlinear process should be investigated for the two-phase flow. Such approach was demonstrated, for example, by Iafrati [51].

The problem of breaking parameterization includes two points: (1) establishment of a criterion of the breaking onset and (2) development of the algorithm of the breaking parameterization. The problem of breaking is discussed in details in Ref. [47]. It was found in Ref. [52] that the clear predictor of breaking formulated in dynamical and geometrical terms, probably, does not exist. The consideration of the exact criterion for the breaking onset for the models using transformation of the coordinate type of (1) is useless since the numerical instability in such models occurs not because of the approach of breaking but because of the appearance of the high local steepness. The description of breaking in the direct wave modeling should satisfy the following conditions: (1) it should prevent the onset of instability at each point of millions of grid points over many thousands of time steps; (2) it should describe in a more or less realistic way the loss of the kinetic and potential energies with preservation of balance between them; and (3) it should preserve the volume. It was suggested by Chalikov [53] that an acceptable scheme can be based on a local highly selective diffusion operator with a special diffusion coefficient. Several schemes of such type were validated, and finally, the following scheme was chosen:

$$\eta_\tau = E_\eta + J^{-1}\left(\frac{\partial}{\partial \xi} B_\xi \frac{\partial \eta}{\partial \xi} + \frac{\partial}{\partial \vartheta} B_\vartheta \frac{\partial \eta}{\partial \vartheta}\right), \tag{23}$$

$$\varphi_\tau = F_\varphi + J^{-1}\left(\frac{\partial}{\partial \xi} B_\xi \frac{\partial \varphi}{\partial \xi} + \frac{\partial}{\partial \vartheta} B_\vartheta \frac{\partial \varphi}{\partial \vartheta}\right), \tag{24}$$

where F_η and F_φ are the right-hand sides of Eqs. (4) and (5) including the tail dissipation terms; B_ξ and B_ϑ are the diffusion coefficients. The probability of high negative values of the curvilinearity is by orders larger than the probability calculated over the ensemble of linear modes with the spectra generated by the nonlinear model.

The curvilinearity turned out to be very sensitive to the shape of surface. This is why it was chosen as a criterion of the approaching breaking. The coefficients B_ξ and B_ϑ depend nonlinearly on the curvilinearity

$$B_\xi = \begin{cases} C_B \eta_{\xi\xi}^2 & \eta_{\xi\xi} < \eta_{\xi\xi}^{cr} \\ 0 & \eta_{\xi\xi} \geq \eta_{\xi\xi}^{cr} \end{cases} \tag{25}$$

$$B_\vartheta = \begin{cases} C_B \eta_{\vartheta\vartheta}^2 & \eta_{\vartheta\vartheta} < \eta_{\xi\xi}^{cr} \\ 0 & \eta_{\vartheta\vartheta} \geq \eta_{\xi\xi}^{cr} \end{cases} \tag{26}$$

where the coefficients at $C_B = 0.05$, $\eta_{\xi\xi}^{cr} = \eta_{\vartheta\vartheta}^{cr} = -50$. The algorithm (24)–(27) does not change the volume and decreases the local potential and kinetic energies. It is assumed that the lost momentum and energy are transferred to the current and turbulence (see Ref. [42]). Besides, the energy also goes to other wave modes. The choice of parameters in Refs. (24)–(27) is based on simple considerations: the local piece of surface can closely approach the critical curvilinearity but not exceed it. The values of the coefficients were chosen in the course of multiple experiments to provide agreement with the rate of spectrum development given by JONSWAP approximation.

5. Evolution of wave field

The integration was done for 1,200,000 steps with the time step $\Delta t = 0.005$ up to the nondimensional time $T = 6000$, which corresponded to 9550 initial wave peak periods. The total energy of wave motion $E = E_p + E_k$ (E_p is the potential energy, while E_k is the kinetic energy) is calculated with the following formulas:

$$E_p = 0.25\overline{\eta^2}, \quad E_k = 0.5\overline{\overline{\left(\varphi_x^2 + \varphi_y^2 + \varphi_z^2\right)}}, \quad E = E_p + E_k, \tag{27}$$

where a single bar denotes the averaging over the ξ and ϑ coordinates, while a double bar denotes the averaging over the entire volume. The derivatives in Ref.

(27) are calculated according to the transformation (1). An equation of the integral energy E evolution can be represented in the following form:

$$\frac{dE}{dt} = \overline{I} + \overline{D_b} + \overline{D_t} + \overline{N}, \tag{28}$$

where $\overline{I}$ is the integral input of energy from wind (Eqs. (14)–(20)); $\overline{D_b}$ is a rate of the energy dissipation due to wave breaking (Eqs. (23)–(26)); $\overline{D_t}$ is a rate of the energy dissipation due to filtration of high-wave number modes ('tail dissipation,' Eqs. (21) and (22)); $\overline{N}$ is the integral effect of the nonlinear interactions described by the right-hand side of the equations when the surface pressure p is equal to zero. The differential forms for calculation of the energy transformations can be, in principle, derived from Eqs. (4)–(6), but here a more convenient and simple method was applied. Different rates of the integral energy transformations can be calculated with the help of fictitious time steps (i.e., apart from the basic calculations). For example, the value of $\overline{I}$ is calculated by the following relation:

$$\overline{\overline{I}} = \frac{1}{\Delta t}\left(\overline{\overline{E^{t+\Delta t}}} - \overline{\overline{E^{t}}}\right), \tag{29}$$

where $\overline{\overline{E^{t+\Delta t}}}$ is the integral energy of a wave field obtained after one time step with the right side of Eq. (6) containing only the surface pressure calculated with Eqs. (14)–(18).

The evolution of the characteristics calculated by formula (29) is shown in **Figure 2**.

The sharp variation of all the characteristics at $t < 500$ is explained by adjustment of the linear initial fields to the nonlinearity. The integral effect of the nonlinear interaction $\overline{\overline{I}}$ (straight line 1) was very close to zero. The tail dissipation $\overline{\overline{D_t}}$ (curve 2) is smaller than the breaking dissipation $\overline{\overline{D_t}}$(curve 3). The value of $\overline{\overline{D_b}}$ has significant fluctuations due to introduction of the criteria (25) and (26). The dissipation $\overline{\overline{D}}_b + \overline{\overline{D}}_t$ absorbs nearly all of the incoming energy, and just a small part of it is going for growth of waves. The balance of energy $\overline{\overline{B}} = \overline{\overline{I}} + \overline{\overline{D}}_t + \overline{\overline{D}}_b$ (curve 5) fluctuates and approaches zero when energy $\overline{\overline{E}}$ (curve 6 in **Figure 2**) approaches saturation.

The time evolution of the integral spectral characteristics is presented in **Figure 3**.

Curve 1 corresponds to the weighted frequency ω_w

$$\omega_w = \left(\frac{\int \omega S dk dl}{\int S dk dl}\right)^{1/2}, \tag{30}$$

where integrals are taken over the entire Fourier domain. The value ω_w is not sensitive to the details of spectrum; hence, it well characterizes the position of spectrum and spectral peak shifting. Curve 2 describes the evolution of the spectral maximum. The step shape of curve corresponds to the fundamental property of downshifting. Opposite to common views, the development of spectrum occurs not monotonically, but by appearance of a new maximum at a lower wave number as well as by attenuation of the previous maximum. It is interesting to note that the same phenomenon is also observed in the spectral model [36].

The value of fetch in the periodic problem can be calculated by integration of the peak phase velocity $c_p = |k|^{-1/2}$ over time.

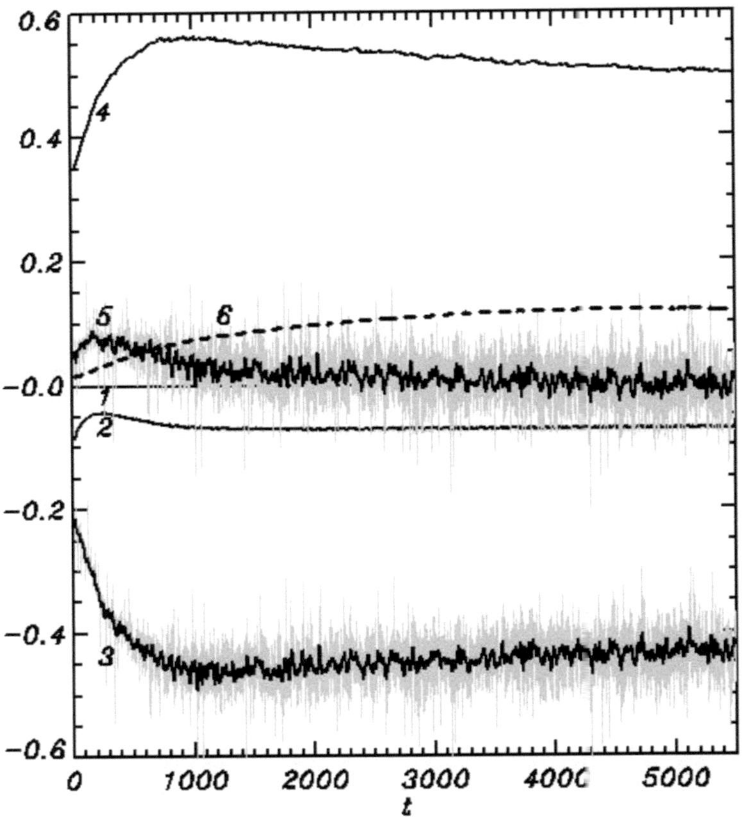

Figure 2.
The evolution of integral characteristics. The rate of evolution of the integral energy multiplied by 10^8 due to: (1) nonlinear interaction $\bar{\bar{I}}$ (Eq. (29)); (2) tail dissipation $\overline{\overline{D}}_t$(Eqs. (21) and (22)); (3) breaking dissipation D_b (Eqs. (23)–(26)); (4) input of energy from wind $\bar{\bar{I}}$ (Eqs. (14)–(20)); and (5) balance of energy $\bar{\bar{I}} + \overline{\overline{D}}_t + \overline{\overline{D}}_b$. Curve 6 shows the evolution of wave energy 10^5E. Vertical gray bars show instantaneous values; thick curve shows the smoothed behavior.

$$F = \int_{t_0}^{t} c_p dt. \tag{31}$$

The numerical experiment reproduces the case when development of wave field occurs under the action of a permanent and uniform wind. This case corresponds to the JONSWAP experiment [29]. It is suggested that the frequency of spectral peak changes as $F^{-1/3}$, while the full energy grows linearly with F Neither of the dependences can be exact since they do not take into account approaching a stationary regime. Besides, the dependence of frequency on fetch is singular at $F = 0$. A more accurate is the approximation:

$$\omega_p = \frac{75.6}{5.63 + F^{1/3}}. \tag{32}$$

Obviously, the dependence $\omega_p \sim F^{-1/3}$ is valid in a narrower interval of F. As seen, contrary to ω_w, the peak frequency changes not monotonically, but by appearance of a new maximum at a lower wave number as well as by attenuation of the previous maximum. It is interesting to note that the same phenomenon is also observed in the spectral model (16). The dependence of the total energy E on fetch F does not look like a linear one, but it is worth to note that the JONSWAP dependence is evidently inapplicable to a very small and large fetch.

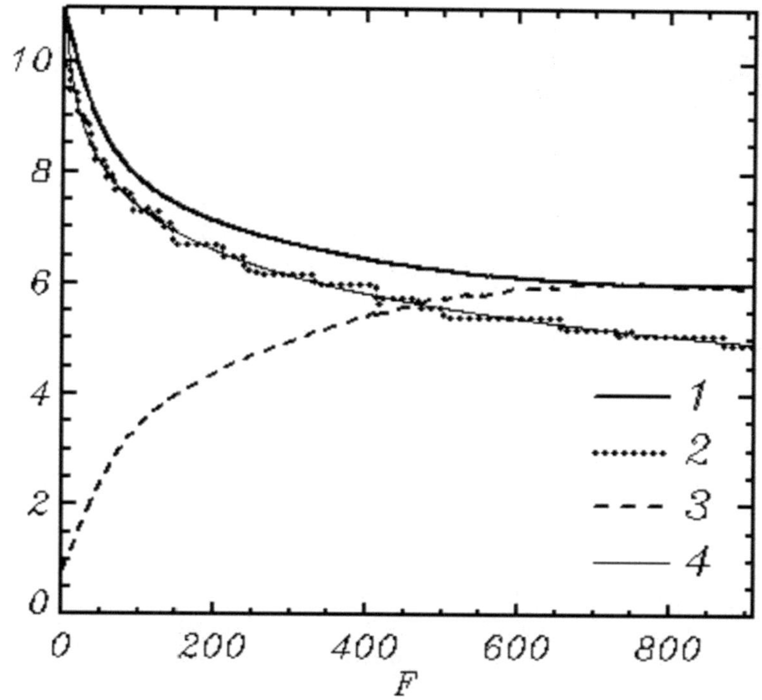

Figure 3.
Dependence of integral characteristics on fetch (Eq. (32)): (1) weighted mean frequency ω_w (Eq. (31)); (2) peak frequency ω_p; (3) energy E (Eq. (27)); and (4) approximation of the peak frequency evolution (Eq. (32)).

On the whole, the evolutions of integral characteristics of the solution shown in **Figures 2** and **3** are smoother than those calculated by Chalikov [26]. It can be explained by multiple technical improvements of the numerical scheme and higher resolution.

The evolution of wave spectrum is shown in **Figure 4**.

The 2-D wave spectrum $S(k,l)$ $(0 \le k \le M_x, -M_y \le l \le M_y)$ averaged over nine time intervals of length equal to $\Delta t \approx 500$ was transferred to the polar coordinates $S_p(\psi,r)(-\pi/2 \le \psi \le \pi/2, \ 0 \le r \le M_x)$ and then averaged over the angle ψ to obtain the 1-D spectrum$S_h(r)$:

$$S_h(r) = \sum S_p(\psi,r) r \Delta\psi. \tag{33}$$

The angle $\psi = 0$ coincides with the direction of wind U, $\Delta\psi = \pi/180$. Even the averaged over angle spectrum looks quite irregular and contains multiple holes and peaks. The spectra are smoothed.

The two-dimensional wave spectra are shown in **Figure 5,** where the $\log_{10}(S(\psi,r))$ averaged over the successive eight periods of length $\Delta t = 500$ is given.

The first panel with a mark 0 refers to the initial conditions. The pictures well characterize the downshifting and angle spreading of spectrum due to the nonlinear interactions.

As seen, each spectrum consists of separated peaks and holes. This phenomenon was observed and discussed by Chalikov et al. [7]. The same results were obtained by Chalikov [26]. The repeated calculations with different resolutions showed that such structure of 2-D spectrum is typical. The locations of peaks cannot be explained by the fixed combination of interacting modes, since in different runs (with the same initial conditions but a different set of phases for the modes), the peaks are located in different locations in the Fourier space.

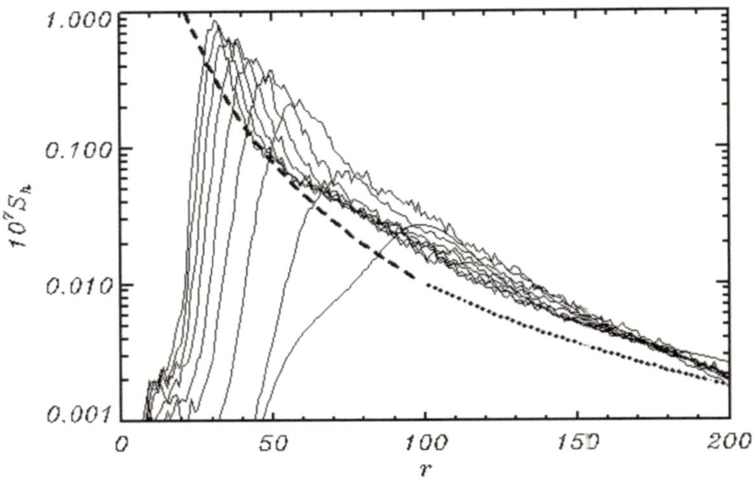

Figure 4.
The wave spectra $S_h(r)$ integrated over angle ψ in the polar coordinates and averaged over the consequent intervals of length about 500 units of the nondimensional time t (thin curves). The spectra are growing and shifting from right to left. Thick dashed curve is the dependence $S \sim \omega^{-4}$; dotted curve corresponds to $S \sim \omega^{-5}$.

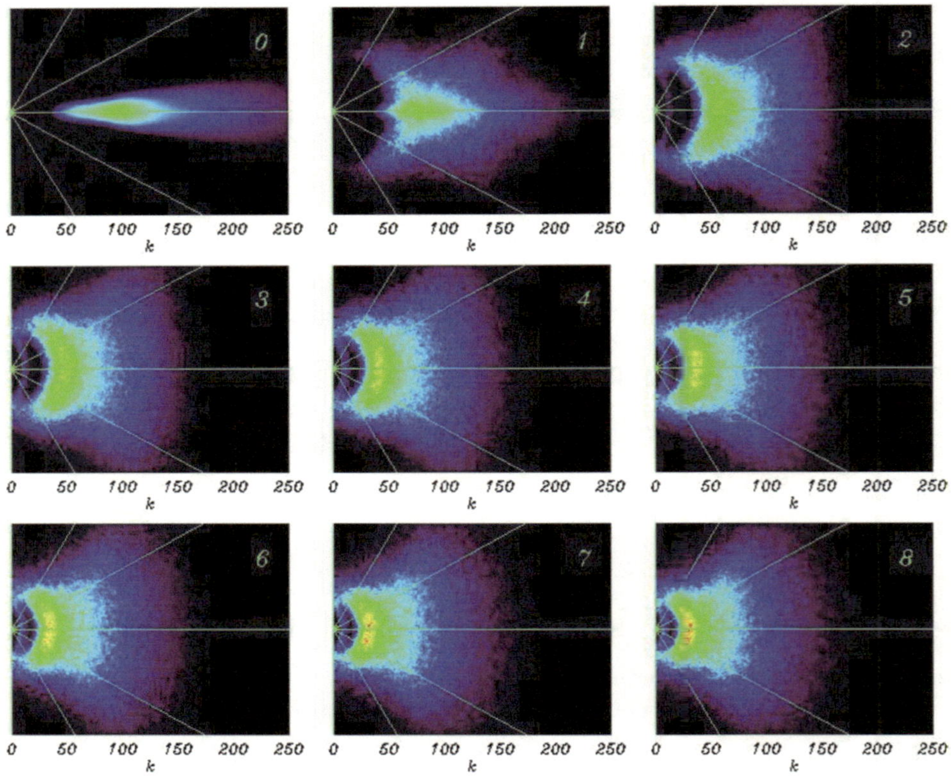

Figure 5.
Sequence of 2-D images of $\log_{10}(S(r, \psi))$ averaged over the consequent eight periods of length $\Delta t = 500$. The numbers indicate the period of averaging (the first panel marked 0 refers to the initial conditions). The spectra are normalized by the maximum value of spectrum 8. The horizontal axis corresponds to the wave numbers $r = |k|$; the angles are shown by rays.

It is interesting to note that while increasing resolution, the patches with low energy extend. It can be supposed that the current and higher resolutions are excessive, and the process can be simulated with a lower resolution. This statement may be too optimistic, but it can be supported by the following arguments. The multi-mode wave mechanics is different from the multi-scale turbulent motion. The modeling of turbulence at increase of resolution just allows reproducing more details of motion. The increase of resolution in a wave model introduces other wave modes with different phase velocities. Due to dispersion, the solutions (i.e., the evolution of surface) in these two cases will be completely different. It means that the solution does not converge with increase of resolution, which makes no sense.

The situation can be saved if upon reaching the optimal resolution, the new added positions for the modes will not obtain the energy and not participate in solution. The existence of such effect should be carefully validated with the exact wave model. If this effect does not exist, it means that the results of simulations depend completely on the resolution, the reliable simulation of individual evolution of wave field being, in fact, impossible.

The method of calculation of the simulated one-dimensional input and dissipation spectra was described by Chalikov [26]; still, it will be explained here once again, though briefly.

The evolution of the integrated over angle ψ wave spectrum $S_h(r)$ can be described with the equation:

$$\frac{dS_h(r)}{dt} = I(r) + D_t(r) + D_b(r) + N(r), \tag{34}$$

where $I(r), D_t(r), D_b(r)$ and $N(r)$ are the spectra of the input energy, tail dissipation, breaking dissipation, and the rate of nonlinear interactions. All of the spectra shown below were obtained by transformation of the 2-D spectra into a polar coordinate (ψ, r) and then integrated over the angles ψ within the interval $(-\pi/2, \pi/2)$. The spectra can be calculated using an algorithm similar to Eq. (29) for integral characteristics. For example, the spectrum of the energy input $I(k,l)$ is calculated as follows:

$$I(k,l) = \left(S_c^{t+\Delta t}(k,l) - S_c^t(k,l)\right)/\Delta t, \tag{35}$$

where $S_c(k_x, k_y)$ is a spectrum of the columnar energy calculated by the relation:

$$S_c(k,l) = \frac{1}{2}\left(h_{k,l}^2 + h_{-k,-l}^2 + \int_{-H}^{0} \left(u_{k,l}^2 + u_{-k,-l}^2 + v_{k,l}^2 + v_{-k,-l}^2 + w_{k,l}^2 + w_{-k,-l}^2\right) d\zeta\right) \tag{36}$$

where the grid values of velocity components u, v, w are calculated by the relations:

$$u = \varphi_\xi + \varphi_\zeta \eta_\xi, \quad v = \varphi_\vartheta + \varphi_\zeta \eta_\vartheta, \quad w = \varphi_\zeta, \tag{37}$$

and $u_{k,l}, v_{k,l}$ and $w_{k,l}$ are the real Fourier coefficients, while for the negative indices—the imaginary ones.

For calculation of $I(k,l)$, the fictitious time steps Δt are made only with a term responsible for the energy input, that is, the surface pressure p. The spectrum $I(k,l)$ was averaged over the periods $\Delta t \approx 500$, then transformed into a polar coordinate system and integrated in the Fourier space over the angles ψ within the interval $(-\pi/2, \pi/2)$. Such procedure was used for calculation of all the terms in the right side of Eq. (34). In the current version of the model, the calculations of integral (28) and spectral (34) transformations were combined with the calculations of the right sides of Eqs. (4) and (5).

The rates of transformation of spectrum are shown in **Figure 6**. The integral term describing the nonlinear interaction $\overline{N}$ in Eq. (28) is small (as compared with the local values of $N_{k,l}$), but the magnitude of spectrum $N(r)$ is comparable with the input $I(r)$ and dissipation $D_t(r)$ and $D_b(r)$ terms (panel 1 in **Figure 5**). The shape of spectrum $N(r)$ confirms prediction of the quasi-linear theory [54, 55]. At the low wave number slope of the spectrum, the nonlinear influx of energy is positive, while at the opposite slope, it is negative. This process produces shifting of spectrum to the lower wave number (downshifting). The input of energy due to the nonlinear interactions is observed in a high frequency part of spectrum, which also agrees with Hasselmann's theory. Note that the nonlinear interactions also produce widening of spectrum.

The spectral distribution of the energy input from wind $I(r)$ (panel 2 in **Figure 6**) is in general similar to wave spectrum since it depends linearly on the spectral density (**Figure 3**). The dissipation rate $D_b(r)$ is negative (panel 3), and its minimum is shifted a little to higher frequencies from the wave spectrum peak. The tail dissipation (Panel 4) is smaller by two orders than the other terms, but it plays an important role of supporting numerical stability.

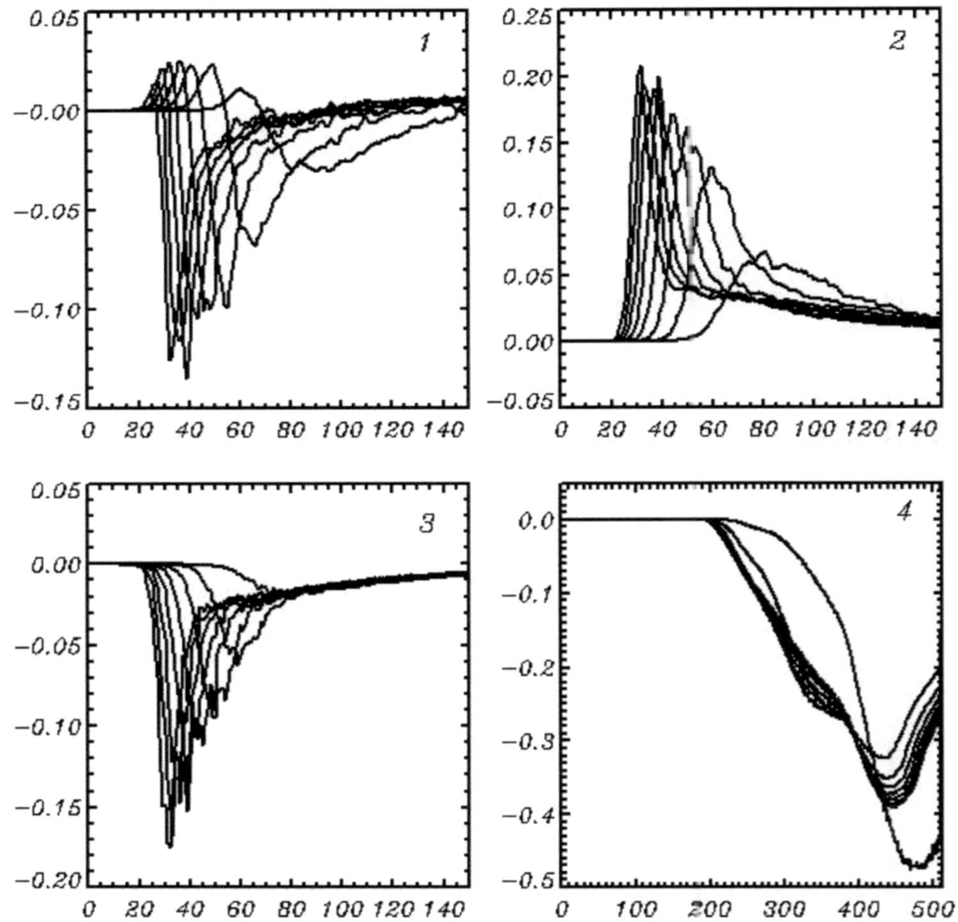

Figure 6.
The rates of transformation spectra multiplied by 10^9: (1) nonlinear interaction $N(r)$ (1); (2) input energy $I(r)$; (3) breaking dissipation $D_b(r)$; (4) tail dissipations $D_t(r)$ multiplied by 10^{11}. All spectra are obtained by transformation of the 2-D spectra into the polar coordinate (ψ, r) and then integrated over the angles ψ within the interval $(-\pi/2, \pi/2)$.

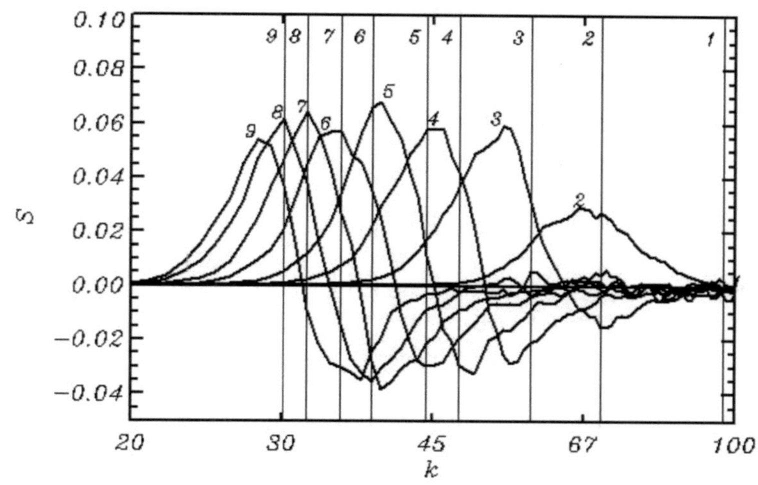

Figure 7.
The spectral distribution of the residual input of energy $dS_h(r)/dt$ in the energy-containing part of spectrum for different stages of wave development. The numbered vertical lines indicate positions of the spectral peak. The second set of numbers shows the peaks of the corresponding residual input.

The residual rate of transformation of spectrum $dS_h(r)/dt$ averaged over eight consequent periods is shown in **Figure 7**. The numbers in the top part of panel indicate the averaged wave number of the spectral peak. The second set of numbers refers to the corresponding spectrum of the residual input of energy. As seen, the maximum of the input energy is located to the left of the spectral peak, that is, on a low-wavenumber spectral slope. The obtained energy causes downshifting of spectrum and supports the shape of a high wavenumber slope and spectral tail. In the equilibrium regime, all the incoming energy are consumed for supporting the shape of the entire spectrum.

However, the dynamics of the tale is not adiabatic, that is, it is not completely controlled by the spectral energy cascade since the input of energy due to the

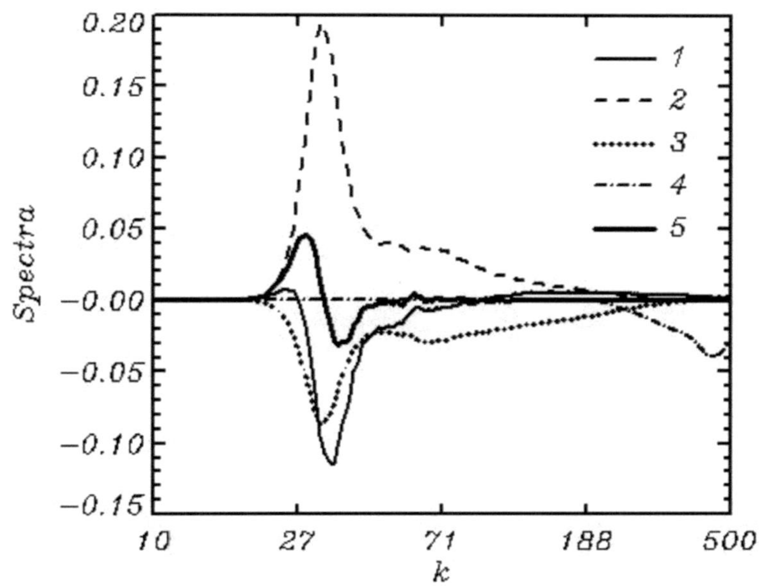

Figure 8.
The rates of transformation spectra multiplied by 10^9 for the last period: (1) nonlinear interaction $N(r)$ (1); (2) input energy; (3) breaking dissipation $D_b(r)$; (4) tail dissipations $10^{10}D_t(r)$; and (5) balance of all the terms.

nonlinear interaction competes with the energy input from wind (curve 2) and breaking dissipation (curves 3 and 4). The tail dissipation (curve 4) is small and concentrated in the vicinity of the cut wave number. The input and dissipation in the spectral tail are nearly in balance (curve 4) (**Figure 8**).

6. Statistical properties of wave field

The phase-resolving modeling requires a higher computer capacity for calculations of any statistical characteristics of sea waves. In the course of simulations, 1.200 two-dimensional fields of the elevation and surface potential with the size 2048 × 1024 points were recorded. Following the solution of the 3-D equation for the velocity potential, these data allow us to reproduce any kinematic and dynamic characteristics of the three-dimensional structure of waves.

The most important statistical characteristics of wave field are mean H_s, variance V, skewness Sk, and kurtosis Ku calculated by the averaging over each of 1200 fields:

$$V = \overline{(\eta - \overline{\eta})^2}, \quad Sk = \overline{(\eta - \overline{\eta})^3} V^{-3/2}, \quad Ku == \overline{(\eta - \overline{\eta})^4} V^{-2} - 3. \tag{38}$$

The evolution of these characteristics in time is shown in **Figure 9**.

The volume of the domain characterized by $\overline{\eta}$ is preserved with the accuracy of the order of 10^{-8}. The variance V is the potential energy that is growing up to the saturation. When the wave field is a superposition of a large number of linear waves, both the skewness and kurtosis are equal to zero. The skewness S characterizes asymmetry of the probability distribution indicating that the positive values of η are larger than the negative ones, then $S > 0$. The kurtosis Ku is positive if the crests are sharper and the troughs are smoother than in the case of linear waves.

The probabilities of the geometrical characteristics (elevation, first and second derivatives over x) are shown in **Figure 10**. The elevation Z (normalized by the significant wave height) is characterized by asymmetry: the heights of waves are significantly larger than the depths of troughs, that is, the wave field is closer to the superposition of Stokes waves than to that of the harmonic modes. The distribution of slopes exhibits horizontal asymmetry: the negative slopes are larger than the

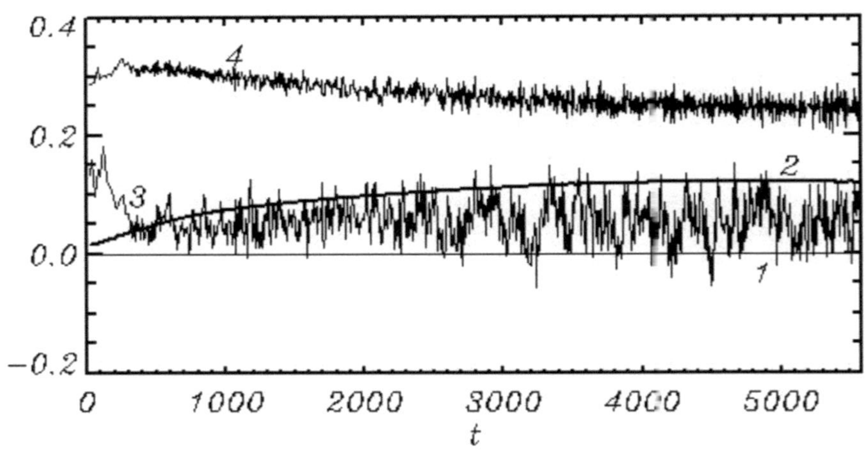

Figure 9.
Evolution of the statistical characteristics of elevation: (1) mean value $\overline{\eta}$, (2) variance V, (3) skewness Sk*, and (4) kurtosis* Ku.

positive ones, that is, the waves, on the average, are inclined in the direction of movements. The second derivative (curvilinearity) has the most striking tendency for asymmetry: the negative values corresponding to the sharpness of crests are much larger by absolute value than the positive values corresponding to the curvilinearity of troughs. This property of curvilinearity was used for the parameterizing of breaking. The limit value $Z_{xx} = -50$ was used as a criterion for the initiating of breaking (see Eq. (26)).

The probability for three components of the surface velocity is given in **Figure 11**. The distributions of the vertical and transverse components of velocity are symmetrical. For a horizontal component, the values of positive fluctuations are considerably larger than the negative fluctuations. This effect cannot be explained by the influence of Stokes drift, which value for those specific conditions does not exceed 10^{-3}. The asymmetry of the probability distribution for the u-components is definitely connected with the asymmetry of the probability distribution for inclinations of surface (**Figure 10**, panel 2).

The number of extreme waves with a high crest $Z_c/H_s > 1.2$ is shown in **Figure 12**. Because such wave is not presented in each of the wave fields, the picture looks as discrete bars of different heights. The total number of values $Z/H_s > 1.2$ is 17.214. The formally calculated probability of the values equals $0.67 \cdot 10^{-5}$. Note that the data on the probability of wave height contain uncertainty because it is not always clear which event should be considered as a single freak wave. The straightforward way suggests calculation of a portion of all the records including freak waves, out of the total volume of the data.

However, some of the records can belong to the single moving freak waves. The cause of this uncertainty is the absence of a strict definition of freak wave being either a case or a process. The number of extreme waves grows with development of wave field.

The integral probability of the total wave height Z_{tc}/H_s, the wave height above mean level Z_c/H_s, and the depth of trough Z_t/H_s are shown in **Figure 13**.

Thin lines show that $Z_{tc}/H_s = 2$ correspond approximately to $Z_c/H_s = 1.2$ and $-Z_t/H_s = -.86$. It is worth to remind that here the nondimensional 'extreme' waves are considered. The true extreme waves are the product of the real wave field. The probability of real extreme waves can be estimated by multiplying the probability of the nondimensional wave by the probability of significant wave height.

The statistical connection between the total wave heights, crest heights, and trough depths is shown in **Figure 14**.

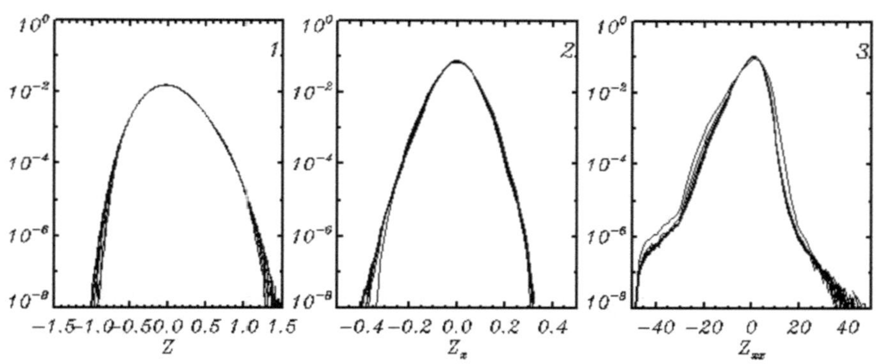

Figure 10.
Geometric characteristics of elevation: (1) probability of elevation $P(Z)$; (2) probability of slopes $P(Z_x)$; and (3) probability of curvilinearity $P(Z_{xx})$.

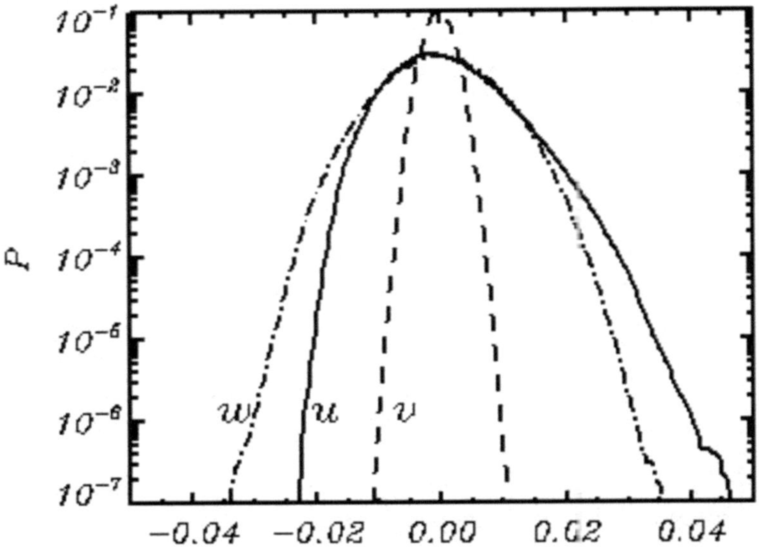

Figure 11.
Probability of the longitudinal u, transverse v, and vertical w components of the surface velocity calculated for the last of nine periods corresponding approximately to the quasi-stationary regime.

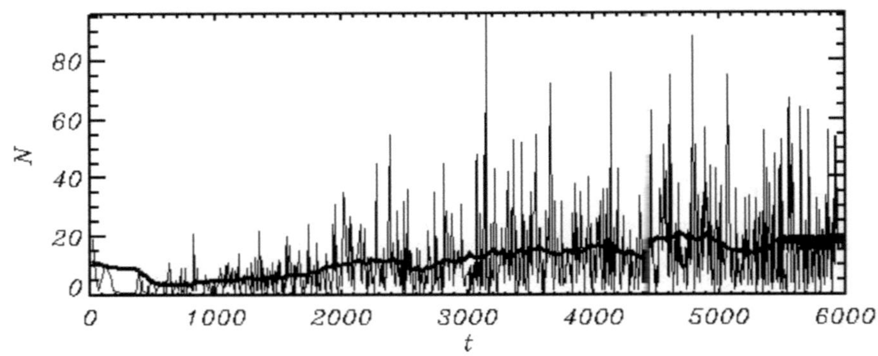

Figure 12.
The number of points where the nondimensional height Z_c/H_s exceeds 1.2. The total number of values is $2.57 \cdot 10^9$.

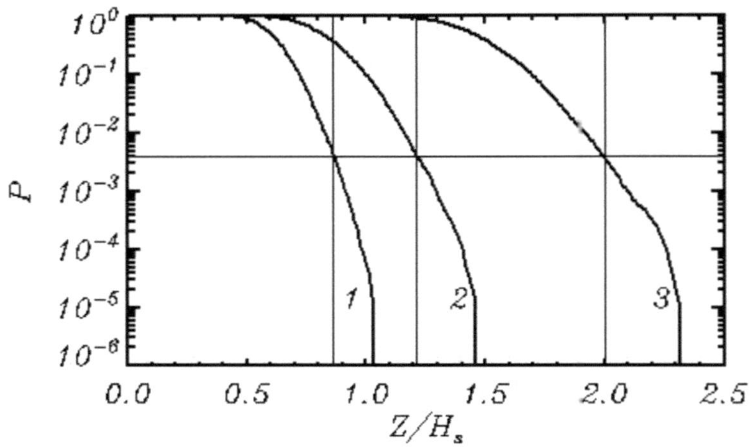

Figure 13.
The cumulative probability of crest-to-trough wave height Z_{tc}/H_s (curve 3); crest height Z_c/H_s (curve 2); and trough depth $-Z_t/H_s$ (curve 3). The number of points in each filed is equal to 2048 × 1024. *The number of fields is* 1200.

The dependences between these characteristics can be approximated by the formulas:

$$\begin{aligned} \tilde{Z}_c &= \text{-}0.105 + 0.626\tilde{Z}_{tc} + 0.015\tilde{Z}_{tc}^2 \\ \tilde{Z}_t &= \text{-}0.105\text{-}0.374\tilde{Z}_{tc} + 0.015\tilde{Z}_{tc}^2 \end{aligned} \tag{39}$$

where the tilde denotes the normalizing by significant wave height H_s. Note that the first and third coefficients in (39) turned out to be a match. The correlation coefficient between $\tilde{Z}_t$ and $\tilde{Z}_c$ is -0.354, while between $\tilde{Z}_t$ and $\tilde{Z}_{tc}$, it is -0.721, and between $\tilde{Z}_c$ and $\tilde{Z}_{tc}$, it is 0.903, that is, the correlation between the full wave height $\tilde{Z}_{tc}$ and the wave height above mean level $\tilde{Z}_c$ is so high that $\tilde{Z}_c$ can be used for identification of extreme waves.

The last characteristics that we consider here is the angle distribution of the spectral density. This characteristic can be described by the function $\Upsilon(\omega/\omega_p)$ (see Ref. [60]).

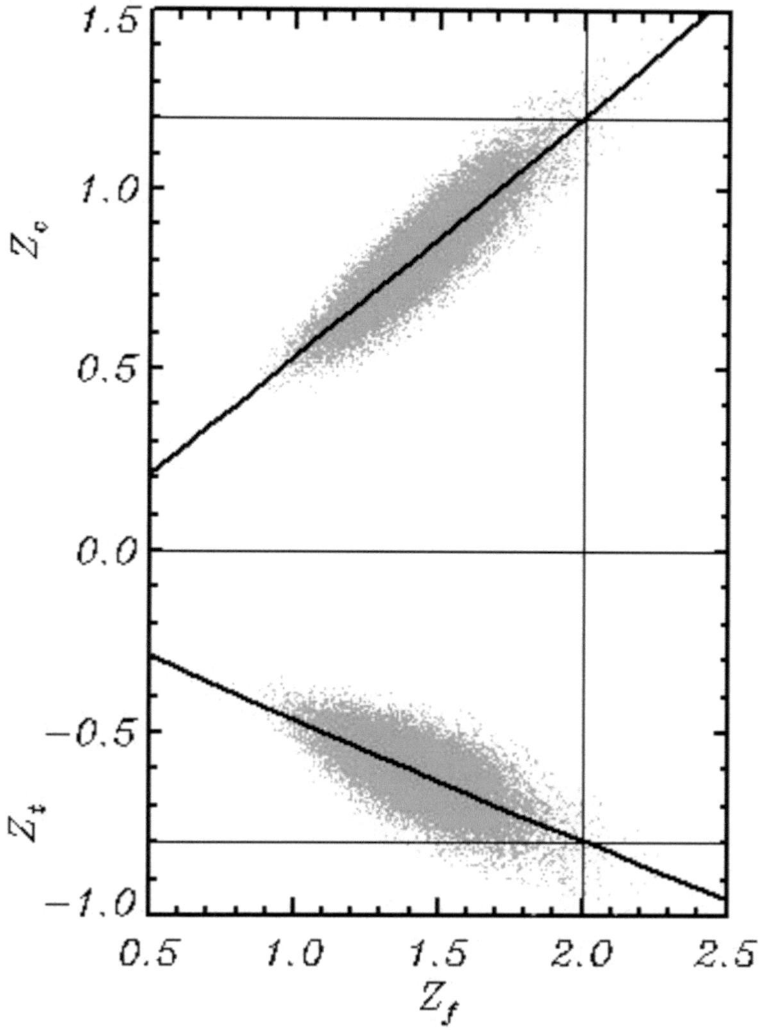

Figure 14.
Dependence of crest height Z_c/H_s (top section) and depth trough Z_t/H_s(bottom section) on the total wave height Z_{tc}/H_s.

$$\Upsilon = \frac{\int S(\omega,\psi)|\theta| d\omega d\psi}{\int S(\omega,\psi) d\omega d\psi} \tag{40}$$

where the integrals are taken over the domain $\{(0 < \omega < \omega_c), (-\pi/2 < \psi < \pi/2)\}$. The value Υ is weighted by the absolute spectrum value of wave direction. The wave spectra as the functions of frequency ω normalized by peak frequency ω_p for the first seven periods are shown in the upper panel of **Figure 15**.

The function $\Upsilon(\omega/\omega_p)$ calculated for the same spectra is given in the bottom panel. As seen, the Υ curves corresponding to different wave ages are close to each other. All of them have a sharp maximum at the frequencies below the spectral peak, a well-pronounced minimum in the spectral peak, and a relatively slow

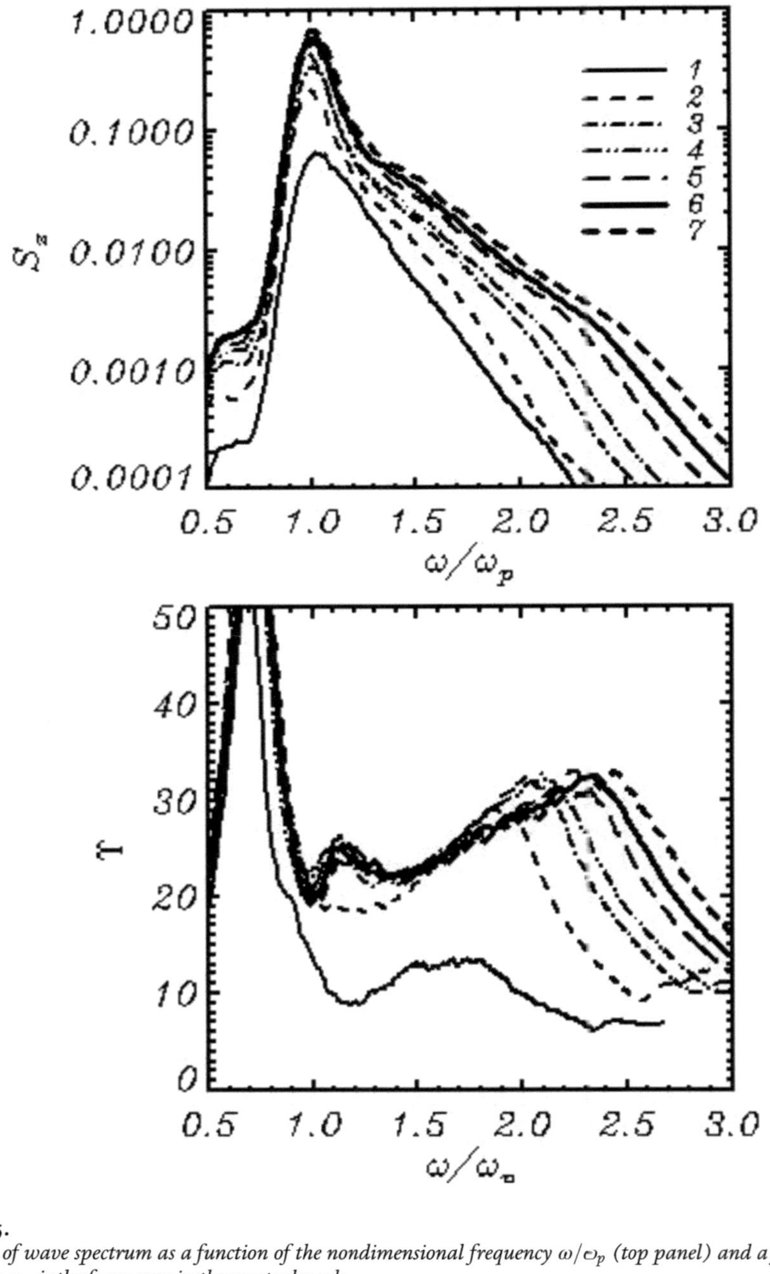

Figure 15.
The shape of wave spectrum as a function of the nondimensional frequency ω/ω_p (top panel) and a function Υ (Eq. (40); ω_p is the frequency in the spectral peak.

growth above the spectral peak. The decrease of Υ at high frequencies is probably caused by the high-frequency dumping. The angle distribution was investigated in Refs. [56–60]. The approximations of $\Upsilon(\omega/\omega_p)$ from the different sources collected in Ref. [61] show considerable scatter, but the general features are quite similar to those calculated in the current work. Note that the spectrum has undergone a long development; hence, the characteristics presented in **Figure 14** were produced by the numerical model itself.

7. Conclusions

The paper is devoted to the wind wave simulations based on the initial equations of potential motion of fluid with a free surface. The system of equations includes the evolutionary kinematic and dynamic surface conditions and Laplace equation for the velocity potential. In this paper, a case of the double-periodic domain of infinite depth is considered. The construction of the exact numerical scheme for a long-term integration of these equations in the Cartesian coordinate system is impossible, since the surface moves between the grid knots. Instead, the system of the curvilinear coordinates (1) fitted with the surface is introduced. The main advantage of this coordinate system is that the surface coincides with a coordinate line $\zeta = 0$. The penalty follows immediately after turning the simple Cartesian coordinates into the curvilinear, nonstationary, and nonorthogonal coordinate system. Fortunately, the evolutionary Eqs. (4) and (5) become just slightly complicated, while Laplace equation transforms into the full elliptic equation. At each time step, these equations can be represented as Poisson equation with the right-hand side depending on the solution itself as well as on the metric coefficient. Since the norm of the right sight of the equation is usually small, the solution of Poisson equation can be found with the three-diagonal matrix algorithm and with iterations over the right-hand side. This procedure being formulated in the Fourier space is greatly simplified by the assumption of periodicity since in this case the derivatives over the horizontal coordinates are represented by the absolute value of wave number $|k|$ in the diagonal terms. When constructing a numerical scheme, we noticed that the significant simplification of the problem can be achieved by separation of the velocity potential into the linear and nonlinear components (see Ref. [7]). It is assumed that the linear component satisfies Laplace equations with the known solution. The equation for the nonlinear component can be obtained by extracting Laplace equation from the initial Poisson equation. Such procedure has a lot of advantages since the nonlinear component is on the average less by 1–2 decimal orders than the linear one. It means that for solution of the reduced Poisson equation the lesser number of levels in vertical, the lesser number of iterations and a smaller accuracy criterion can be used. The use of two components in the evolutionary equation does not seem to provide noticeable advantages; however, this way deserves further consideration.

The adiabatic version of the model was validated by simulation of a running Stokes wave with the steepness $AK = 0.40$ in Ref. [7]. It was shown that the amplitudes of Stokes modes remain practically constant up to the accuracy of 10^{-7}. The current version of the model after some technical improvements of the numerical scheme provides accuracy up to 10^{-12}. Then, the adiabatic version of the model was used for reproduction of a quasi-stationary regime for investigation of the statistical properties of sea waves [1, 7, 8].

For calculations of development of wave field under the action of wind, it was necessary to include the algorithms for calculations of input and dissipation of energy. The scheme for calculation of the energy input was developed by Chalikov and Rainchik [3] on the basis of coupling the one-dimensional phase-resolving model and the two-dimensional boundary layer model with the second-order turbulence closure scheme. The parameterization suggested is still quasi-linear (similar to Miles' scheme [30]), but in our opinion, it is the only scheme confirmed by the extended results of the numerical simulations. The theoretical and observational data on β−function are dramatically scattered (see Ref. [4], **Figure 1**).

For stabilization of the solution, the algorithm of high-frequency dumping in the Fourier space suggested by Chalikov and Sheinin [28] was used. The numerous attempts were made to improve that scheme (for example by reduction of the spectral interval of dumping) but without much success.

The most complicated problem is the parameterization of dissipation due to wave breaking. Such algorithm should not describe a process of breaking as it is, which within the frame of such model is impossible, but it should prevent the numerical instability that interrupts a run (see discussion in Ref. [26]). Currently, the algorithm used is very simple. It is based on the diffusion operator with a highly selective coefficient of 'viscosity.' It works satisfactorily, but we are far from thinking that it cannot be substantially improved or completely replaced by another one.

The results described in this paper show that the wave field development under the action of wind is reproduced quite realistically. The area of application of such models is very wide. Such modeling should be used for improvement of the algorithms of the energy input and dissipation. A model with the periodic boundary conditions can be used for the local interpretation of the spectral forecast in terms of real waves. The finite-difference version of the model can be used for simulation of wave regimes in the basins with real shapes and bathymetry (see, e.g., Ref. [5]).

Acknowledgements

Author would like to thank Mrs. O. Chalikova for her assistance in preparation of the manuscript. This research was performed in the framework of the state assignment of Russian Academy of Science (Theme No. 0149-2019-0015) supported in part 15 (Section 2) by RFBR (project No. 18-05-01122).

Author details

Dmitry Chalikov[1,2]

1 Shirshov Institute of Oceanology RAS, Moscow, Russia

2 University of Melbourne, Victoria, Australia

*Address all correspondence to: dmitry-chalikov@yandex.ru

References

[1] Chalikov D, Babanin AV. Comparison of linear and nonlinear extreme wave statistics. Acta Oceanologica Cinica. 2016;**5**(5):99-105. DOI: 10.1007/313131-016-0862-5

[2] Troitskaya YI, Sergeev DA, Kandaurov AA, Baidakov GA, Vdovin MA, Kazakov VI. Laboratory and theoretical modeling of air-sea momentum transfer under severe wind conditions. Journal of Geophysical Research. 2012;**117**:C00J21. DOI: 10.1029/2011JC007778

[3] Chalikov D, Rainchik S. Coupled numerical modelling of wind and waves and the theory of the wave boundary layer. Boundary-Layer Meteorol. 2011; **138**:1-41. DOI: 10.1007/s10546-010-9543-7

[4] Chalikov D, Babanin AV. Parameterization of wave boundary layer. Atmosphere. 2019;**2020**(10):686. DOI: 10.3390/atmos10110686

[5] Engsig-Karup AP, Harry B, Bingham HB, Lindberg O. An efficient flexible-order model for 3D nonlinear water waves. Journal of Computational Physics. 2009;**228**(6):2100-2118

[6] Engsig-Karup A, Madsen M, Glimberg S. A massively parallel GPU-accelerated mode for analysis of fully nonlinear free surface waves. International Journal for Numerical Methods in Fluids. 2012. DOI: 10.1002/fld. 2675

[7] Chalikov D, Babanin AV, Sanina E. Numerical modeling of three-dimensional fully nonlinear potential periodic waves. Ocean Dynamics. 2014; **64**(10):1469-1486. DOI: 10.1007/s10236-014-0755-0

[8] Chalikov D. Numerical Modeling of Sea Waves. Springer; 2016. p. 330. DOI: 10.1007/978-3-319-32916-1

[9] Causon DM, Mingham CG, Qian L. Developments in multi-fluid finite volume free surface capturing methods. Advances in Numerical Simulation of Nonlinear Water Waves. 2010:397-427

[10] Ma Q, Yan W, Qale S. FEM method and its application to the simulation of free responses of floating bodies and overturning waves. Advances in Numerical Simulation of Nonlinear Water Waves. 2010:165-202

[11] Greaves D. Application of the finite volume method to the simulation of nonlinear water waves. Advances in Numerical Simulation of Nonlinear Water Waves. 2010:357-396

[12] Grue J, Fructus D. Model for fully nonlinear ocean wave simulations derived using Fourier inversion of integral equations in 3D. Advances in Numerical Simulation of Nonlinear Water Waves. 2010:1-42

[13] Ducrozet G, Bonnefoy F, Le Touzé D, Ferrant P. 3-D HOS simulations of extreme waves in open seas. Natural Hazards and Earth System Sciences. 2007;7:109-122. DOI: 10.5194/nhess-7-109-2007

[14] Ducrozet G, Bingham HB, Engsig-Karup AP, Bonnefoy F, Ferrant P. A comparative study of two fast nonlinear free-surface water wave models. International Journal for Numerical Methods in Fluids. 2012;**69**:1818-1834

[15] Ducrozet G, Bonnefoy F, Le Touzé D, Ferrant P. HOS-Ocean: Open-source solver for nonlinear waves in open ocean based on high-order spectral method. Computer Physics Communications. 2016. DOI: 10.1016/j.cpc.2016.02.017

[16] Touboul J, Kharif C. Two-dimensional direct numerical

simulations of the dynamics of rogue waves under wind action. Advances in Numerical Simulation of Nonlinear Water Waves. 2010:43-74

[17] Bonnefoy F, Ducrozet G, Le Touzé D, Ferrant P. Time-domain simulation of nonlinear water waves using spectral methods. In advances in numerical simulation of nonlinear water waves. Advances in Coastal and Ocean Engineering. 2010;**11**:129-164. DOI: 10.1142/9789812836502_0004

[18] Dalrymple RA, Gómez-Gesteira M, Rogers BD, Panizzo A, Zou S, Crespo AJ, et al. Smoothed particle hydrodynamics for water waves. Advances in Numerical Simulation of Nonlinear Water Waves. 2010;**2010**;465-495

[19] Issa R, Violeau D, Lee E-S, Flament H. Modelling nonlinear water waves with RANS and LES SPH models. Advances in Numerical Simulation of Nonlinear Water Waves. 2010:497-537

[20] Lubin P, Caltagirone J-P. Large eddy simulation of the hydrodynamics generated by breaking waves. Advances in Numerical Simulation of Nonlinear Water Waves. 2010:575-604

[21] Kim KS, Kim MH, Park JC. Development of MPS (moving particle simulation) method for multi-liquid-layer sloshing. Journal of Mathematical Problems in Engineering. 2014;**2014**: Article ID 350165, 13 pages. DOI: 10.1155/2014/350165

[22] Zhao X, Liu B-J, Liang S-X, Sun Z-C. Constrained interpolation profile (CIP) method and its application. Chuan Bo Li Xue/Journal of Ship Mechanics. 2016; **20**:393-402. DOI: 10.3969/j.issn.1007-7294.2016.04.002

[23] Young D-L, Wu N-J, Tsay T-K. Method of fundamental solutions for fully nonlinear water waves. Advances in Numerical Simulation of Nonlinear Water Waves. 2010:325-355

[24] Chalikov D, Babanin AV. Simulation of one-dimensional evolution of wind waves in a deep water. Physics of Fluid. 2014;**26**(9): 096607

[25] McLean JW Instabilities of finite-amplitude water waves. Journal of Fluid Mechanics. 1982;**114**:315

[26] Chalikov D. Numerical modeling of surface wave development under the action of wind. Ocean Science. 2018;**14**: 453-470. DOI: https://doi.org/10.5194/os-14-453-2018

[27] Thomas LH Elliptic Problems in Linear Differential Equations over a Network. New York: Columbia University; 1949

[28] Chalikov D, Sheinin D. Direct modeling of one-dimensional nonlinear potential waves. Nonlinear ocean waves. In: Perrie W, editor. Advances in Fluid Mechanics. Vol. 17. 1998. pp. 207-258

[29] Hasselmann K, Barnett RP, Bouws E, et al. Measurements of wind-wave growth and swell decay during the Joint Sea Wave Project (JONSWAP). Tsch. Hydrogh. Z. Suppl. 1973;**A8**(12): 1-95

[30] Miles JW. On the generation of surface waves by shear flows. Journal of Fluid Mechanics. 1957;**3**:02. DOI: 10.1017/S0022112057000567

[31] Snyder RL, Dobson FW, Elliott JA, Long RB. Array measurements of atmospheric pressure fluctuations above surface gravity waves. Journal of Fluid Mechanics. 1981;**102**:1-59

[32] Hsiao SV, Shemdin OH. Measurements of wind velocity and pressure with a wave follower during MARSEN. Journal of Geophysical Research. 1983;**88**:9841-9849

[33] Hasselmann DJ, Bösenberg J. Field measurements of wave-induced

pressure over wind-sea and swell. Journal of Fluid Mechanics. 1991;**230**: 391-428

[34] Donelan MA, Babanin AV, Young IR, Banner ML, McCormick C. Wave follower field measurements of the wind input spectral function. Part I. Measurements and calibrations. Journal of Atmospheric and Oceanic Technology. 2005;**22**:799-813

[35] Donelan MA, Babanin AV, Young IR, Banner ML. Wave follower field measurements of the wind input spectral function. Part II. Parameterization of the wind input. Journal of Physical Oceanography. 2006;**36**:1672-1688

[36] Rogers WE, Babanin AV, Wang DW. Observation-consistent input and whitecapping-dissipation in a model for wind-generated surface waves: Description and simple calculations. Journal of Atmospheric and Oceanic Technology. 2012;**29**(9):1329-1346

[37] Gent PR, Taylor PA. A numerical model of the air flow above water waves. Journal of Fluid Mechanics. 1976; **77**:105-128

[38] Al'Zanaidi MA, Hui HW. Turbulent airflow over water waves—A numerical study. Journal of Fluid Mechanics. 1984; **148**:225-246

[39] Chalikov DV. Numerical simulation of wind-wave interaction. Journal of Fluid Mechanics. 1978;**87**:561-582

[40] Chalikov DV. Numerical simulation of the boundary layer above waves. Boundary-Layer Meteorology. 1986;**34**: 63-98

[41] Chalikov D, Makin V. Models of the wave boundary layer. Boundary-Layer Meteorology. 1991;**56**:83-99

[42] Chalikov D, Belevich M. One-dimensional theory of the wave boundary layer. Boundary-Layer Meteorology. 1992;**63**:65-96

[43] Chalikov D. The parameterization of the wave boundary layer. Journal of Physical Oceanography. 1995;**25**: 1335-1349

[44] Tolman H, Chalikov D. On the source terms in a third-generation wind wave model. Journal of Physical Oceanography. 1996;**26**:2497-2518

[45] Tolman HL and the WAVEWATCH III °R Development Group: User Manual and System Documentation of WAVEWATCH III °R Version 4.18 Environmental Modeling Center Marine Modeling and Analysis Branch. 2014; Contribution No. 316

[46] Sullivan PP, McWilliams JC. Dynamics of winds and currents coupled to surface waves. Annual Review of Fluid Mechanics. 2010;**42**:19-42

[47] Babanin AV. Breaking and Dissipation of Ocean Surface Waves. Cambridge University Press; 2011. p. 480

[48] Alves JHGM, Banner ML. Performance of a saturation-based dissipation-rate source term in Modeling the fetch-limited evolution of wind waves. Journal of Physical Oceanography. 2003;**33**:1274-1298

[49] Babanin AV, Tsagareli KN, Young IR, Walker DJ. Numerical investigation of spectral evolution of wind waves. Part II: Dissipation term and evolution tests. Journal of Physical Oceanography. 2010;**40**(4):667-683

[50] Chabchoub A, Monty J, Nelli F, Lee J, Elsnab J, Toffoli A. An experimental comparison of velocities underneath focused breaking waves. Ocean Engineering. 2018;**155**:201-210

[51] Iafrati A. Numerical study of the effects of the breaking intensity on wave

breaking flows. Journal of Fluid Mechanics. 2009;**622**:371-411

[52] Chalikov D, Babanin AV. Simulation of wave breaking in one-dimensional spectral environment. Journal of Physical Oceanography. 2012;**42**(11): 1745-1761

[53] Chalikov D. Statistical properties of nonlinear one-dimensional wave fields. Nonlinear Processes in Geophysics. 2005;**12**:1-19

[54] Hasselmann K. On the non-linear energy transfer in a gravity wave spectrum, part 1. Journal of Fluid Mechanics. 1962;**12**:481-500

[55] Hasselmann SK, Hasselmann JH, Allender HG, Barnett TP. Computations and parameterizations of the nonlinear energy transfer in a gravity-wave spectrum. Part II: Parameterizations of the nonlinear energy transfer for application in wave models. Journal of Physical Oceanography. 1985;**15**(11): 1378-1392

[56] Donelan MA, Hamilton J, Hui WH. Directional spectra of wind-generated waves. Philosophical Transactions of the Royal Society. 1985;**315A**:509-562. DOI: 10.1098/rsta.1985.0054

[57] Ewans KC. Observations of the directional spectrum of fetch-limited waves. Journal of Physical Oceanography. 1998;**28**:495-512. DOI: 10.1175/1520-0485(r1998)028,0495: OOTDSO.2.0.CO;2

[58] Babanin AV, Soloviev YP. Variability of directional spectra of wind-generated waves, studied by means of wave staff arrays. Marine and Freshwater Research. 1998;**49**:89-101. DOI: 10.1071/MF96126

[59] Banner ML. Equilibrium spectra of wind waves. Journal of Physical Oceanography. 1990;**20**:966-984. DOI: 10.1175/1520-0485(1990)020,0966: ESOWW.2.0.CO;2

[60] Romero L, Melville WK. Airborne observations of fetch-limited waves in the Gulf of Tehuantepec. Journal of Physical Oceanography. 2010;**40**: 441-465. DOI: 10.1175/2009JPO4127.1

[61] Liu Q, Rogers WE, Babanin AV, Young IR, Romero L, Zieger S, et al. Observation-based source terms in the third-generation wave model WAVEWATCH III: Updates and verification. Journal of Physical Oceanography. 2019;**49**(2):489-517. DOI: 10.1175/JPO-D-18-0137.1

Chapter 6

Surface Gravity Wave Modeling in Tropical Cyclones

Yalin Fan, Paul Hwang and John Yu

Abstract

Tropical cyclones are among the deadliest geophysical phenomena on earth. Tropical cyclone-generated wave fields are of interest both scientifically for understanding wind–wave-ocean interaction physics and operationally for predicting potentially hazardous conditions for ship navigation and coastal regions. This chapter briefly reviews the development of third generation wave models, the improvements of their input/dissipation source functions, and their applications in tropical cyclone generated surface wave predictions. Discussion on the status of coupled atmosphere-wave-ocean modeling in tropical cyclone predictions are given at the end of the chapter prompted by the growing scientific evidence on the importance of sea state on air-sea fluxes under extreme wind conditions.

Keywords: surface gravity wave, tropical cyclone, wave modeling, wave forecast, atmosphere–ocean-wave coupled models, tropical cyclone forecast

1. Introduction

Tropical cyclones, also popularly known as hurricanes or typhoons, are among the most spectacular and deadly geophysical phenomena. Not only the intense winds associated with the storms can create enormous waves, the ocean wave field generated by tropical cyclones are extremely complex with a combination of swell and wind sea due to the spatially inhomogeneous and directionally varying wind fields, and the directional distribution of the wind sea component is often skewed due to the rapid variation in the wind direction. Thus, the tropical cyclone-generated wave fields are of interest not only operationally for predicting potentially hazardous conditions for ship navigation and coastal regions, but also scientifically for understanding wind–wave-ocean interaction physics.

There have been considerable efforts made to understand the characteristics of tropical cyclone-generated surface waves through both measurements and numerical modeling. Several third generation wave models such as WAVEWATCH III [1], the Wave Model (WAM) [2], Simulating Waves Nearshore (SWAN) [3], University of Miami Wave Model (UMWM) [4], etc., have been used to study surface wave responses during hurricanes. The third generation wave models, as well as its predecessors (the first and second generation wave models), are all spectra models. Which means that the model solves the evolution of the surface wave energy spectral instead of the physical form of the surface wave itself, and thus it is also known as non-phase resolving wave model. Although phase resolving wave models have been actively developed during the past two decades, the spectral models are still the only approach capable of solving the temporal and spatial variations in the oceanic surface gravity wave fields. Thus, it will be the focus of this chapter. The readers interested in phase resolving wave modeling can read the chapter on "High-Resolution Numerical Simulation of Surface Wave Development under the Action of Wind" by Dr. Dmitry Chalikov. In this chapter, we will review the progress on the development of the third generation models, their applications to tropical cyclone wave forecasts, the improvements made to the input and dissipation source function in the model, and the challenges we face in advancing the wave forecast skills.

Since tropical cyclones are driven by enthalpy fluxes from the sea and limited mostly by surface drag, being able to accurately estimate the momentum and heat flux under these extreme wind conditions is critical for tropical cyclone predictions. During the past three decades, more and more scientific evidence has suggested that the air-sea fluxes is closely coupled to the sea state in the ocean [5–9]. With the continuous improvements in surface wave forecasts under tropical cyclone conditions, fully coupled atmosphere-wave-ocean model is suggested for accurate hurricane predictions as well as corresponding ocean responses [10–13]. However, the outcome from the coupled models are mixed, which we will discuss in more details at the end of the chapter.

2. Wave forecasts under tropical cyclones

2.1 Early developments in wave modeling

Ocean surface gravity waves are long regarded as a basic parameter of interest for marine engineering and navigation applications. Hence, it is necessary to develop the capability to forecast wave conditions over global and regional ocean domains to minimize loss of life and property. The basis for modern wave research was laid in the 1950s and 1960s. The first computer generated wave forecasts were made in 1956 by the Joint Numerical Weather Prediction Unit (JNWP) at Suitland, Maryland [14], which produced a single wave height and period at each grid point using a simple relationship between the local wind speed and duration and the wave height and period.

An important advance was the introduction of the concept of a wave spectrum by Pierson et al. [15], in which the random wave field is broken into a spectrum of many regular wave components which are distinguished by wavenumber vector (k), and relative or intrinsic frequency (σ). σ is also called angular frequency or radial frequency because it is measured in radian. Another popularly used frequency variable is f, which is measured in hertz (Hz) with $\sigma = 2\pi f$. Later on, an experimental milestone, the Joint North Sea Wave Project (JONSWAP) experiment [16], was conducted. In which, among other things, the fetch dependence of the spectral evolution was observed and the concept of self similarity of the spectral shape emerged. Following the success of the JONSWAP, rapid improvements were made in spectral wave modeling by solving the radiative transfer equation:

$$\frac{dN}{dt} = \frac{S_{in} + S_{nl} + S_{ds} + S_{bot} + \dots}{\sigma} \tag{1}$$

where, $N(k,;,\theta,;,x,;,t) = F(k,;,\theta,;,x,;,t) / \sigma$ is the wave action density spectrum, $F(k,;,\theta,;,x,;,t)$ is the wave number-direction spectrum, θ is the wave direction, x is the vector represents the coordinate system in the geographical space, and t is time. The right side of the equation represents a combination of non-conservative sources and sinks of the wave energy with S_{in} represent the wind input source term, S_{ds} represent the dissipation due to wave breaking, S_{nl} represent the transfer of energy due to nonlinear interactions between the spectral wave components, and S_{bot} stands for dissipation due to bottom friction. Other source terms can be easily added such as surf breaking, bottom scattering or reflection by shoreline or iceberg. They are neglected in Eq. (1) here since they are not the focus of this chapter.

The classification of different spectral models is largely based on the treatment of the nonlinear interaction term (S_{nl}). In the so-called first generation models, S_{nl} is not modeled explicitly, so that all spectral components evolve independently. Dissipation for wind seas is generally modeled as an on–off mechanism, limiting the spectral evolution to some pre-described spectral shape. In second-generation models, simple approximations for nonlinear interactions are introduced, either treating the entire wind sea part of the spectrum using empirical growth relations and idealized spectral shapes (so-called hybrid models), or by modeling S_{nl} based on results for simplified spectral shapes (so-called discrete models).

After the Sea Wave Model Project (SWAMP) study in the mid-1980s, through community efforts, the Wave Model (WAM) was developed to solve Eq. (1) with explicit treatment the S_{nl} term, essentially replacing all previous models and marked the beginning of third-generation wave model era [17, 18]. The WAM was a major step forward in wave modeling, and it has been validated and applied to wave hindcast and Forecast over many seas of the world [19–21]. Since its development, the WAM model has been actively used by many wave research and forecast groups, including the European Center for Medium-range Weather Forecasts (ECMWF).

Despite the success of the WAM model, evaluations carried out at the National Centers for Environmental Prediction (NCEP) suggested that this model also left room for further improvement [22], such as the use of first-order numeric in the propagation terms that adversely influences swell propagations; the large fixed time steps used in source terms integrations can result in spectral shape errors in rapidly changing wave conditions, and extreme conditions were systematically underestimated due to an artifact of the physical parameterizations. WAVEWATCH III [23] was developed at NCEP in the spirit of the WAM. It is designed with more general governing transport equations that permit full coupling with ocean models, improved propagation schemes, improved physics integration scheme, and improved physics of wave growth and decay. It has been validated both over global-scale wave forecast and regional wave forecast [1, 24–26], and it was the first wave model validated for detailed wave spectra simulations under hurricane conditions.

For near-shore applications, Simulating Waves Nearshore (SWAN) model was developed at the Delft University of Technology [27]. Compare to WAM, it includes more flexible options on the parameters for processes such as non-linear wave–wave interactions, wind wave generation, energy dissipation by breaking, and friction and frequency shifting due to current and local topographical conditions. After being satisfactorily verified with field measurements [3, 28], it was the first model used to simulate tropical cyclone waves in the coastal waters of Taiwan Island.

The University of Miami Wave Model (UMWM) was developed [4] aimed at an efficient wave model to provide full atmosphere-wave-ocean coupling in hurricane forecasting systems [29]. Thus, the source functions that drive the space-time evolution of the energy spectra are developed in form based on theory and laboratory and field experiments under extreme wind conditions of tropical cyclones. The calibration factors (proportionality constants of the source functions) are determined from a comparison of modeled and observed significant height and mean period during Hurricane Bonnie (1998) and Hurricane Ike (2008). Although the modeled spectral shapes by UMWM in the four quadrants of Hurricane Bonnie (1998) match the Scanning Radar Altimeter measurements better than other spectral wave models, its overall performance against measurements from varies platforms shows less accuracy [30].

2.2 Wave predictions under tropical cyclones

The first wave modeling study under extreme tropical cyclone conditions was conducted by Ou et al. [31] using SWAN within the coastal waters of Taiwan. Model simulated significant wave height during the passages of four typhoons are compared with measurements at several wave stations near the island. The model results look reasonable on the east coast of the island while large discrepancies are found for the comparisons on the west coast of the island. The authors attribute the large errors to the simple wind field used to force the wave model, which is generated using a parametric model and did not account for the effect of the island's central mountains that partly damage the cyclonic structures of the passing-over typhoons.

While significant wave height is a useful information to have, it only gives a general idea of the total wave energy in the wave group. The directional wave spectrum contains information of the distribution of wave energy in wave number and direction space, and thus can be used to identify different contributions to local wave energy, e.g. swell from distant storms and locally wind-generated waves. The direction of propagation of wave energy and period ($1/f$ or $2\pi/\sigma$) of the most energetic waves are important for many practical applications, e.g. the design and operation of coastal and offshore structures and storm surge forecasts. Furthermore, the limited point measurements from a few moored buoy stations or oil platforms cannot reflect the spatial patterns of wave fields very well. Thus, considerable efforts have been made to measure the directional spectra of tropic cyclone generated surface waves and to investigate its spectral characteristics. Wyatt [32] described measurements of the directional spectra of storm waves using high frequency radar to explain the effect of fetch on the directional spectrum of Celtic Sea storm waves. Holt et al. [33] examined the capability of synthetic aperture radar imagery from ERS-1 satellite to track the wave fields emanating from an intense storm over a several day period. Wright et al. [34] and Walsh et al. [35] studied the spatial variation of hurricane directional wave spectra for both open ocean and landfall cases using the National Aeronautics and Space Administration (NASA) Scanning Radar Altimeter (SRA) for the first time through a joint effort between the NASA Goddard Space Flight Center and the National Oceanic and Atmospheric Administration (NOAA)/Hurricane Research Division (HRD). These measurements have provided detailed wave characteristics along the flight tracks of the NOAA aircraft carrying the SRA, and many SRA measurements have been carried out during hurricanes in the North Atlantic since.

To evaluate the ability of third generation wave models in prediction of directional wave spectra, Moon et al. [36] simulated Hurricane Bonnie (1998), a category 2–3 tropical cyclone on the Saffir-Simpson hurricane intensity scale (SSHS), when it approached the U.S. East Coast using WAVEWATCH III. Input and dissipation source function package ST2 was chosen for their simulation. Details about this source function package are given in Section 2.3. The results from their simulations are compared with buoy observations and NASA SRA data, which were obtained on 24 August 1998 in the open ocean and on 26 August when the storm was approaching the shore. While the model results yielded good agreement with observations of directional spectrum as well as significant wave height, dominant wavelength, and dominant wave direction (wavelength and direction at the peak frequency of the wave spectrum) excluding shallow areas near the shore, later studies found that WAVEATCH III overestimates the significant wave height under very high wind conditions in strong hurricanes [37–39]. These studies attribute this error to the overestimations of the drag coefficient (C_d) used in the wave model at very high winds.

Powell et al. [5] estimated C_d using a dataset from hundreds of global positioning system (GPS) sondes that were dropped in the vicinity of hurricane eyewalls, where the strongest wind occurs, in both the Atlantic basin and the eastern and central Pacific basins since 1997. This is among the first estimates of C_d in tropical cyclones under high wind speeds over 40 m/s. Their analysis found that surface momentum flux levels off as the wind speed increases above hurricane force, a behavior contradictory to surface flux parameterizations in a variety of modeling applications at the time. Inspired by their study, Donelan et al. [6] further studied the aerodynamic friction between air and sea under extreme winds in laboratory settings. They confirmed that the aerodynamic roughness approaches a limiting value in high winds, and a fluid mechanical explanation of this phenomenon was given based on their study. More comprehensive studies on the air-sea fluxes were carried out later on through the Coupled Boundary Layer Air-Sea Transfer experiment (CBLAST), a cooperative undertaking between the Office of Naval Research (ONR), NOAA's Oceanic and Atmospheric Research (OAR) lab, HRD, Aircraft Operations Center (AOC), including its US Weather Research Program (USWRP), and the U.S. Air Force Reserve Command's 53rd Weather Reconnaissance Squadron "Hurricane Hunters", which yielded an unprecedented dataset for exploring the coupled atmosphere and ocean boundary layers during an active hurricane [40]. Key results from the analysis effort to date have increased the range of air–sea flux measurements significantly, which have allowed drag and enthalpy exchange coefficients to be estimated in wind speeds to nearly hurricane force.

Fan et al. [11] investigated the effect of different drag coefficient parameterizations in WAVEWATCH III through a modification to the input/dissipation source package ST2 using a very strong tropical cyclone, Ivan (2004). Hurricane Ivan (SSHS category 4–5 in the Caribbean Sea and Gulf of Mexico) was one of the most intensively observed hurricane to date. Three sets of detailed SRA wave spectra measurements were collected as well as satellite measurements and National Data Buoy Center (NDBC) buoy time series, providing a nice temporal and spatial coverage along the passage of the hurricane. The illustration of the location of these measurements from their paper is given here in **Figure 1**.

The authors also utilized the NOAA/HRD real-time wind analysis (HWIND) as their model forcing. HWIND is an integrated tropical cyclone observing system in which wind measurements from a variety of observation platforms are used to develop an objective analysis of the distribution of wind speeds in a hurricane [41]. The spatial resolution of HWIND is about 6 km × 6 km and covers an area of about 8° × 8° in latitude–longitude around the hurricane's center. The wind field was usually provided near real time at intervals of every 3 or 6 hours. Although HWIND provides excellent spatial representation of the hurricane wind field, its coarse temporal resolution and small spatial coverage is not sufficient to force a numerical model, and was only used for theoretical wind field analysis after the product became publicly available since 1994.

To take advantage of this wind product, Fan et al. [11] introduced a normalized interpolation technique to interpolate the HWIND field in time and extrapolate it in space with minimum distortion of the hurricane wind field. Results from their wave simulation experiments suggested that the model with the original ST2 drag coefficient parameterization tends to overestimate the significant wave height and the dominant wavelength and produces a wave spectrum with narrower directional spreading. When an improved drag parameterization that considers the level off at

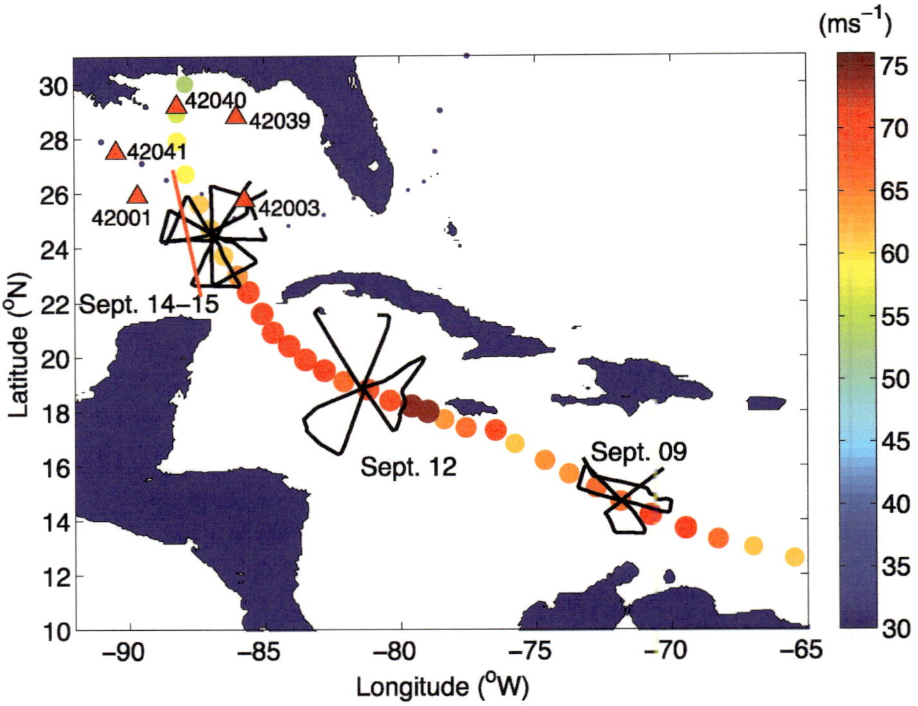

Figure 1.
Available measurements along Hurricane Ivan track. The color and size of the circle represents the maximum wind speed of the hurricane. The black lines in the vicinity of the hurricane track represent the aircraft storm relative flight tracks during the SRA measurements. The red line to the left of the hurricane track overlaps with the September 14–15 SRA measurements shows the satellite tracks of Envisat-1 and ERS-2. The red triangles in the Gulf of Mexico show National Data Buoy Center buoy locations along hurricane Ivan track.

high wind is introduced, the model yields an improved forecast of significant wave height when compared with SRA, satellite, and NDBC buoy measurements, but underestimates the dominant wavelength. The SRA model comparison on Sept 9 from their paper is given here in **Figure 2** as an example to illustrate the improvements in wave height simulations and the bias in wave length simulations. This bias was later on corrected with improved input and dissipation source functions as discussed in Section 2.3 below.

Most importantly, Fan et al. [11] investigated the effect of ocean current inputs on wave predictions in their study and found that the effect of wave-current interaction on hurricane wave predictions are even stronger than the improved C_d (**Figure 2**), especially when the hurricane moves over a preexisting mesoscale ocean feature, such as the Loop Current in the Gulf of Mexico or a warm- and cold-core ring, the current associated with the feature can accelerate or decelerate the wave propagation and significantly modulate the wave spectrum. Detailed idealized experiments conducted in Fan et al. [42] suggested that in the right-forward quadrant of the hurricane center where the currents are strong and roughly aligned with the dominant wave propagating direction, the advection effect of currents can introduce an absolution (relative) error in significant wave height as large as 2 m (~20%).

Since WAVEWATCH III was shown to perform better than SWAN under tropical cyclone conditions [43, 44], and was thus more popularly used by researchers and operational centers for surface wave simulations under extreme wind conditions, our discussion on wave modeling under tropical cyclone conditions will focus on WAVEWATCH III from hereafter.

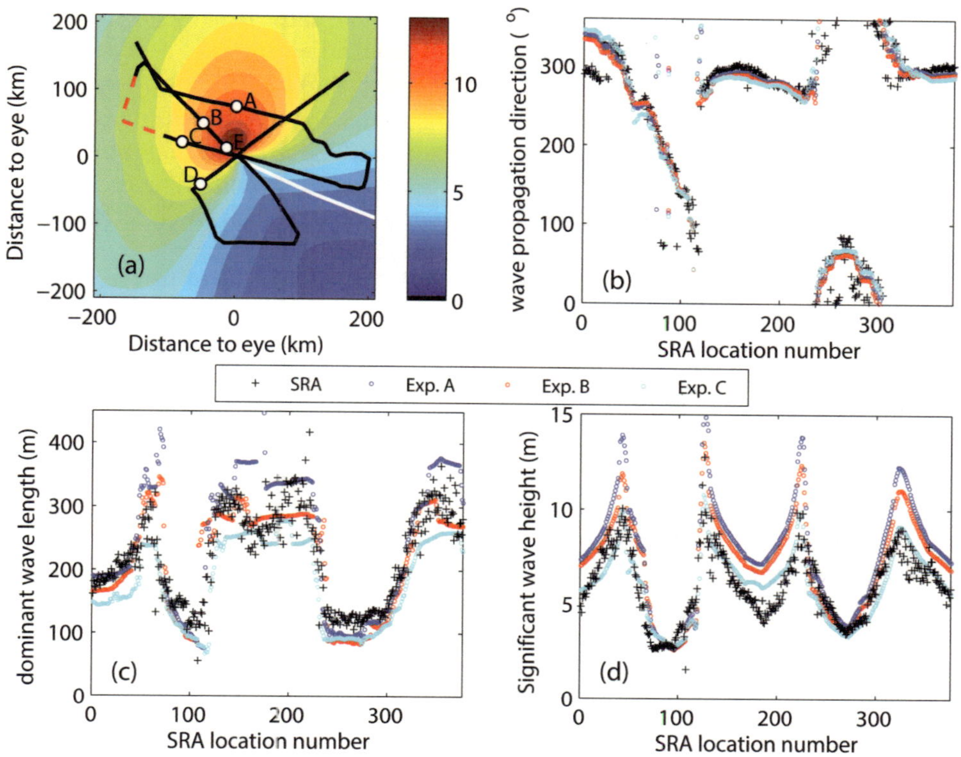

Figure 2.
(a) Significant wave height field (m, color) at 1800UTC on 9 September. The thick white line is the hurricane track and thick gray line is the flight track. The black arrow shows the start point and direction of the flight, and the black dots shows the SRA location in an increment of every 50 data points from the start. (b) Wave propagation direction relative to true north rotating clockwise, (c) dominant wavelength, and (d) significant wave height comparison between SRA measurements and model results in experiments A, B, and C corresponding to simulations using original c_d, modified c_d, and modified c_d plus wave-current interaction respectively.

2.3 Input and dissipation source functions

2.3.1 Developments of input and dissipation source functions in WAVEWATCH III

There are five different input/dissipation source term packages in WAVEWATCH III referred to as ST1, ST2, ST3, ST4, and ST6. Each model describes the wind generation and whitecapping dissipation differently. Generally, the term describing the wind input is determined as:

$$S_{in} = \beta(k,\theta)N(k,\theta)\sigma \qquad (2)$$

where $\beta(k,\theta)$ is the dimensionless wind-wave growth rate parameter.

The $\beta(k,\theta)$ used in the first source package (ST1) is based on the source terms of WAM cycles 1 through 3 [45, 46]. It is an empirical formula as a function of the 10 m wind speed (U_{10}) and direction (θ_w) and the wave phase velocity (c_{ph}) and direction (θ). The drag coefficient C_d in this formulation is defined as a linear function of U_{10}, and a cap (2.5×10^{-3}) is applied to C_d for high winds based on previous findings for hurricane wave simulations.

Source package ST2 is initially developed by Tolman and Chalikov [47] and later on updated by Tolman [48]. It combines a wind input adjusted to the numerical model of airflow above waves by Chalikov and Belevich [49], in which $\beta(k,\theta)$ is a nondimensional wind-wave interaction parameter that varies with C_d and the dimensionless frequency of the spectral components. The wind input terms in ST2 can become negative for waves traveling at large angles with wind or faster than wind, and thus is a better representation of energy flow at the air-sea interface and a big improvement over the ST1 input source term.

Their dissipation term is also improved over ST1 by consisting of two separate terms for both the low frequency waves and the high-frequency tail of the spectrum, whose shape is adjusted to produce a roll-off of the wave spectrum proportional to f^{-5} at high frequencies, as proposed by Phillips [50].

Model results using ST2 has shown significant improvement over that using ST1 by being able to produce excellent growth behavior from extremely short fetches up to full development, giving smoother results and is less sensitive to numerical errors [47].

ST3 adapted the ECMWF WAM parameterization described by Bidlot [51]. This parameterization combines the wind input term originally based on the wave growth theory of Miles [52] with the feedback on the wind profile parameterized by Janssen [53], and the input source function is a function of the wave supported stress τ_w:

$$\tau_w = \left| \int_0^{k_{max}} \int_0^{2\pi} \frac{S_{in}(k',\theta)}{C} (cos\theta, sin\theta) dk' d\theta + \tau_{hf}(u_*, \alpha)(\cos\theta_u, \sin\theta_u) \right| \tag{3}$$

where, k' and θ are the wave number and direction, C is the wave phase speed, θ_u is the wind directions, u_* is the friction velocity, and α is the Charnock coefficient. Eq. (3) for τ_w includes the resolved part of the spectrum, up to the maximum wave number k_{max}, as well as the stress supported by shorter waves, τ_{hf}. Thus, to calculate the roughness parameter, the feedback of the wind-waves spectra is taken into account as well. So, in the considered parameterizations, the wind input is determined by the wind-wave interaction parameter as well as the friction velocity u_*.

This model added a linear swell dissipation component introduced by Janssen [54] to represent the shear stress variations in phase with the orbital velocity, and the mean frequency also occurs in the definition of the maximum frequency of prognostic integration of the source terms. A limitation of their dissipation source function is that it is too sensitive to swell. An increase in swell height typically reduces dissipation at the wind-sea peak, and increase dissipation at high frequencies.

Both ST4 and ST6 inherited the wind input source function from ST3, and focused on the improvement on the dissipation source function in the model.

The least understood aspect of the physics of wave evolution is the dissipation source function. Following Hasselmann's [55] idea that white capping is the main cause for the dissipation process and local in space, Phillips [56] argues that wave dissipation is rather local in wavenumber space. This is followed by Jenkins [57] who advocated the picture that breaking waves will generate ocean eddies (turbulence) that will damp the waves. During the next two to three decades, several dissipation source functions have been proposed and widely used in third generation wave models such as [46, 47]. However, these parameterizations were adjusted to close the wave energy balance instead of using the quantitative relationship with observed feathers. Following the pioneering work by Banner and Young [58], Banner et al. [59, 60] have analyzed breaking in relationship to the formation and related instabilities of groups. Babanin et al. [61], Babanin et al. [62], and Ardhuin et al. [63] worked on the physics of the process analyzing both laboratory and open-field data. These efforts led to new insights into the process of whitecapping, in a way making even more evident the limits associated with the various parameterizations in use. Ardhuin et al. [63] is the first to implement these findings into an operational wave model (WAVEWATCH III, ST4) through a dissipation function without any prescribed spectral shape but based on the empirical knowledge of the breaking of random waves from previous researches and the dissipation of swells over long distances due to air friction. Their work is immediately followed by Babanin [64] and Zieger et al. [65] who implemented the ST6 package in WAVEWATCH III that argues the swell attenuation is due to the interaction with ocean turbulence, and thus swells will transfer energy into the ocean when they dissipate rather than to the air.

2.3.2 Evaluation of different source functions in tropical cyclones

These input/dissipation source packages are evaluated in many studies and for different locations and scenarios. Using Hurricane Ivan (2004) as the test case, Liu et al. [30] conducted the first comprehensive evaluation of the relative strengths/weaknesses of all WAVEWATCH III source functions but ST1 under tropical cyclone conditions. Through the comparison of bulk wave parameters (i.e., significant wave height, mean wave direction and period) derived from SRA spectra measurements, satellite observations, and NDBC buoy data, the authors found that ST3, ST4, and ST6 have comparable skills on wave predictions under extreme wind conditions and significant outperformed the ST2 source package. Their comparisons with the SRA data are given in **Figure 3** for illustration. We can see that while ST2 has similar skills as other sources functions on wave direction predictions, it under predicts the significant wave height and mean wave period. One possible explanation for this is because the upper limiter on C_d adopted by ST2 (C_d, max = 2.5×10^{-3}) starts being active when U_{10} is far below the hurricane wind forcing ($U_{10} \sim 15$ ms^{-1}), which will influence the well-tuned wind wave growth behavior under low to moderate winds [47] and may influence the high wave predictions by ST2.

Another important feature to be noticed in **Figure 3** is the underprediction of wave period by ST2, a model bias reported by Fan et al. [11] as well (we can easily relate wave period to wave length through the dispersion relations), while good prediction skills on the wave period are found using ST3/4/6. This has suggested that more physical based new input/dissipation source functions were able to correct this bias efficiently.

Another wave model evaluated in Liu et al. [30] is the University of Miami Wave Model (UMWM) [4]. It was devised as an efficient wave model to provide full atmosphere-wave-ocean coupling in hurricane forecasting systems [29]. Thus, the physics-based but time-consuming nonlinear interaction source term S_{nl} (e.g., [66–68]) was treated parametrically in such a way that wave breaking was assumed to be the primary cause of the shift of energy to the longer waves. Based on the comparisons between model results and measurements from various platforms, such as the comparison with the SAR measurements in **Figure 3**, the authors concluded that UMWM shows less accuracy than WAVEWATCH III in specification of bulk wave parameters. This is possibly because (i) UMWM-estimated drag coefficient does not clearly show a saturation trend when wind

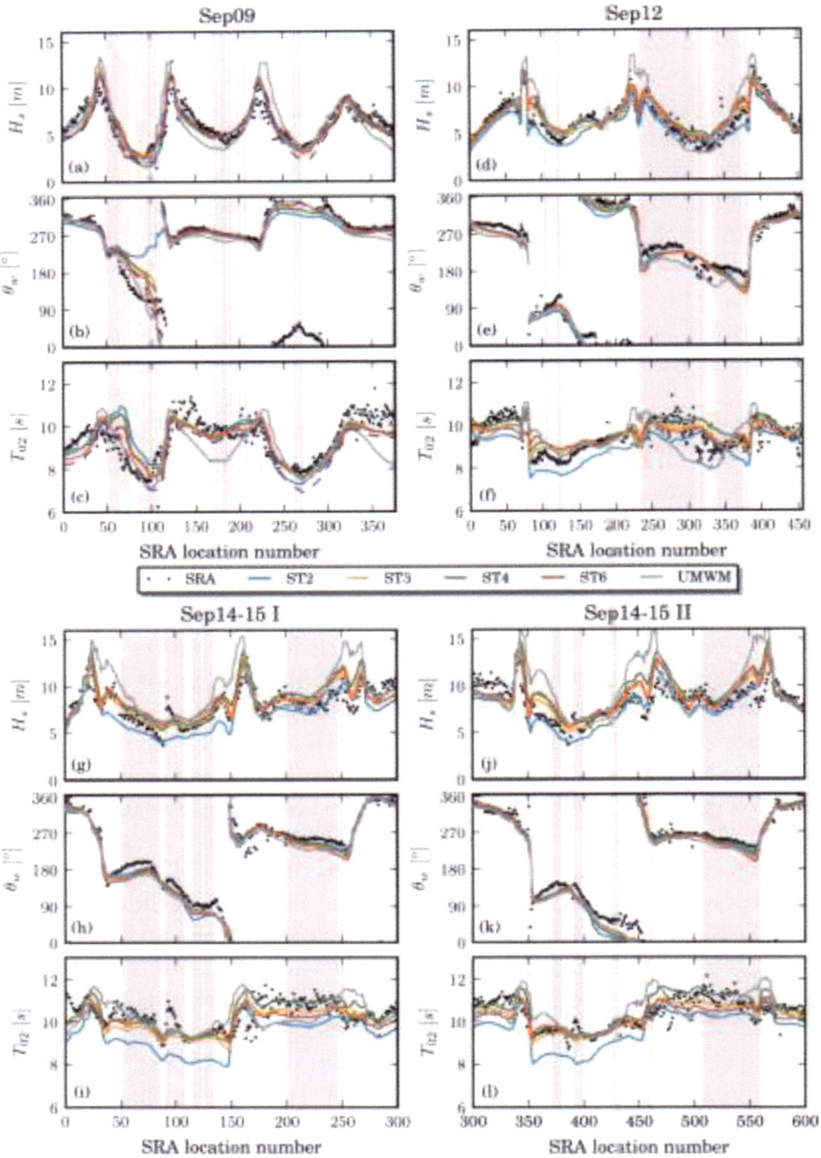

Figure 3.
Comparison of model results (colored lines: Blue for ST2, yellow for ST3, green for ST4, red for ST6 and gray for UMWM) and SRA observations (black dot •) acquired on September (a–c) 9, (d–f) 12 and (g–l) 14–15. For clarity, the SRA measurements on September 14–15 is divided into two parts. One (the first 300 records) is plotted in panels (g–i) and the other (the remaining 300 records) in panels (j–l). Three bulk parameters are taken into account: (a, d, g, j) significant wave height H_s, *(b, e, h, k) mean wave direction* θ_w *(oceanographic convention: the direction towards which waves are propagating, measured clockwise from geographic north) and (c, f, i, l) mean wave period* T_{02}*. The purple dashed lines in panel (a–c) represent the results from the ST6 + WRT experiment.*

speeds are beyond ~35 ms^{-1} and (ii) the four-wave interaction term of UMWM disagrees evidently with the full solution of the Boltzmann integral in detail.

3. Coupled models

As we all know, tropical cyclones are among the deadliest geophysical phenomena. Both the most lethal and the most expensive natural disasters in U.S. history were tropical cyclones [69]. However, being able to accurately predict the intensity and track of these storms is still a big challenge. While the primary driving source for the tropical cyclones are the heat transfer from the ocean through evaporation, the sea surface drag works to slow the storm down. There is little understanding of the behavior of these fluxes at very high wind speeds. Traditionally, due to technical limitations, direct measurements of the fluxes have only been made at wind speeds as large as 25 m/s. As a result, momentum transfer under extreme wind conditions has been extrapolated from these field measurements in a variety of modeling applications, including hurricane risk assessment and prediction of storm motion, intensity, waves and storm surges. However, drop sonde measurements by Powell et al. [5] and laboratory experiments by Donelan et al. [6] suggested that in those extreme circumstances the drag decreases with wind speed or saturates. Their work has opened a new chapter for tropical cyclone prediction models, but the understanding of the physics of such extreme events is only beginning.

Many studies following their pioneer work have suggested that the momentum flux at the air-sea interface is closely coupled with sea state in the ocean. Makin [7] argues that spray production may give rise to the reduction of drag coefficient, C_d, by suppressing the air turbulence for increasing wind speed during hurricanes. On the other hand, Andreas [8] has proposed that when spray returns to the water, short waves will be extinguished. This will no doubt reduce the drag considerably as the short waves carry most of the wave-induced stress [9]. Donelan et al. [6] also suggest that flow separation may be the reason for drag reduction since the outer airflow does not "see" the troughs of the waves during such events and thus unable to follow the wave surface, and skips from breaking crest to breaking crest. All these hypotheses are standing on one common ground – the momentum flux is closely coupled with the sea state in the ocean.

Given the success in wave modeling, there is keen desire in the modeling community to calculate momentum flux using the source function from the wave model and fully coupled Atmosphere-wave-ocean model is suggested for accurate hurricane predictions as well as corresponding ocean responses [11–13]. Although the newly developed fully coupled Atmosphere-wave-ocean models were shown to be able to improve model simulated surface wind and inflow angle in individual storms [13, 70, 71], no statistically significant improvements were observed in intensity forecasts by the Hurricane Forecast Improvement Program [72–75].

The wave energy spectrum computed by the wave models is from a balance between input and dissipation, and the wave parameters that are usually validated against observations are weighted by energy thus depend primarily on long waves around the peak. Since the momentum flux depends mainly on short wind waves, one may ask whether the model spectra represent real spectra well enough to provide reasonable momentum flux to atmosphere and ocean models in a coupled system for tropical cyclone predictions, or is there a stronger argument for using parameterized fluxes? To answer this question, Fan and Rogers [76] compared the drag coefficient computed using WAWEWATCH III simulated wave spectrum under Hurricane Ivan with that calculated using the SRA measured wave spectrum. The Donelan et al. [77] source function was used for these calculations because its stress calculation is based on the wave spectra and wind only, which is suitable for the SRA measurements. In order to quantify the uncertainties brought in by adding spectra tail to the SRA measurements, the model spectra were truncated at the SRA resolution and frequency range and added high frequency tail in the same way. The reconstructed model spectra were shown to give no noticeable difference in wave parameter and drag coefficient calculations from the original spectra.

From their study, the authors found that the drag coefficients disagree between the SRA and model spectra mainly in the right/left rear quadrant of the hurricane

(**Figure 4**) where the observed spectra appear to be bimodal while the model spectra are single peaked with more energy in the swell frequencies and less energy in the wind sea frequencies. The authors also found that the modeled wind sea part, which is essential for stress calculations, is more problematic than the swells. The reason for the large discrepancy in drag coefficients due to the spectra shape differences is because the surface waves are young and vary significantly with time and space in hurricanes, the momentum flux across the air-sea interface under such sea states depends mainly on short wind waves. Since the wind stress is a vector

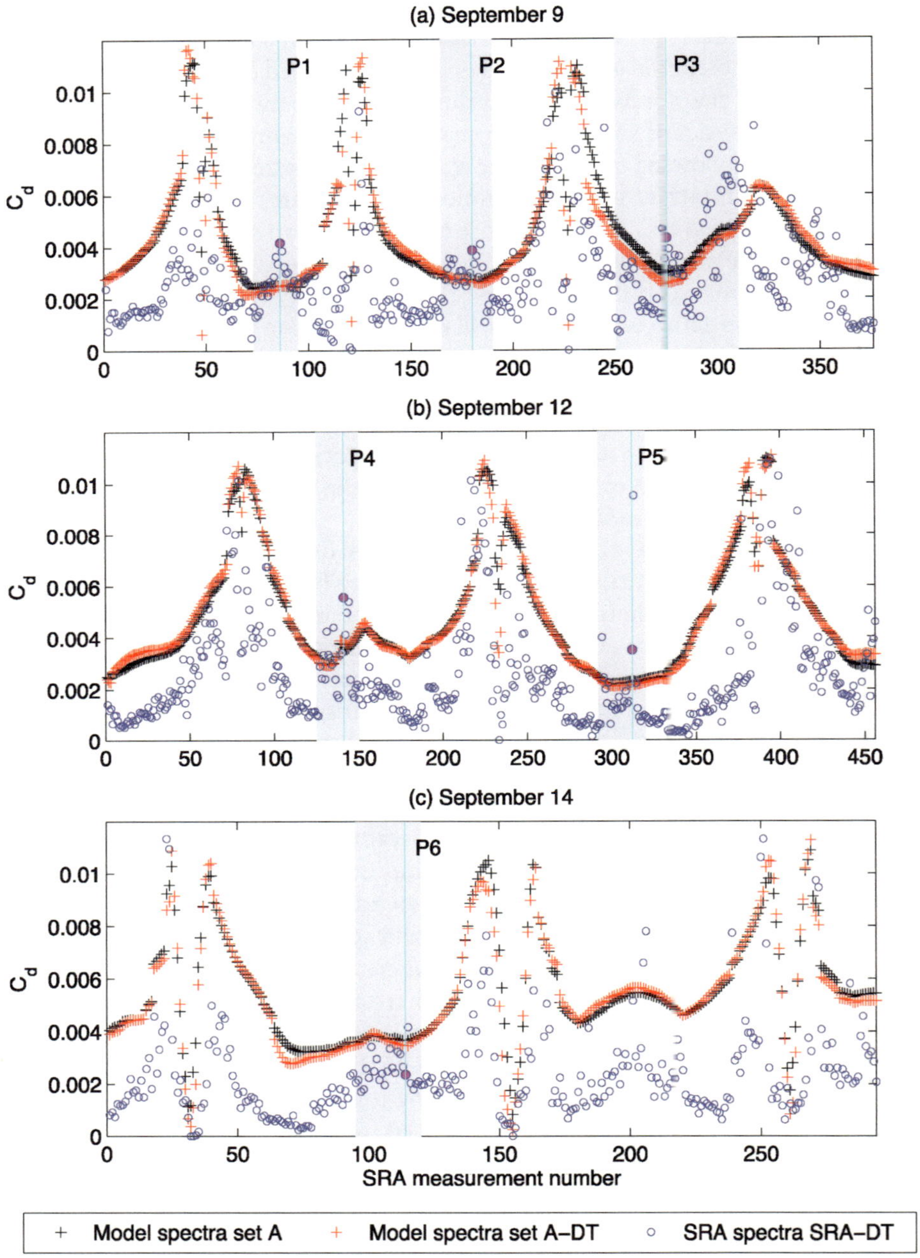

Figure 4.
Drag coefficient (C_d) calculated from original model spectra (set A, black cross), reconstructed model spectra (set A-DT, red cross), and constructed SRA spectra (SRA-DT, blue circle) using the Donelan et al. [77] source function along the flight track of (a) September 9, (b) September 12, and (c) September 14. The gray areas are corresponding to the part of flight track in the right/left rear quadrant of the hurricane.

sum of the momentum contributions in all frequencies and directions of the wave spectrum, the shape of the wave spectra will directly affect its magnitude.

The authors attributed the large discrepancies in the wave spectra simulations to the usage of HRD winds as the wave model forcing. Because the Hwinds are created through temporal and spatial interpolation/averaging of all available observations relative to the storm center, including land, sea, space, and air-borne platforms, the resulting wind field is smoother than real tropical cyclone winds and do not have the fine structures. However, although the wind field produced by the coupled atmosphere-ocean-wave models do have fine structures in the wind field, they failed to produce bimodal wave spectra as well. Which suggests that either the structure of the modeled wind fields is far from reality or there is some other physics missing in the wave model to capture the bimodal waves. For that, Fan and Rogers [76] commented on the dissipation terms in the wave model being developed with no any specific attention to the unique conditions of tropical cyclones. They may generate too little dissipation for the swells and too much dissipation for the wind sea under high wind conditions. This would result in modeled wind sea part of the spectra are continuous (unimodal) and narrower in frequency space compared to observations.

Since the Donelan et al. [77] wind input source functions used in their study was derived from measurements taken only under low to moderate wind conditions [45, 77], it may not be suitable for high wind conditions such as the hurricanes. Fan and Rogers [76] also conducted alternate drag calculations using Donelan et al. [4] source function that was developed for hurricane conditions. They found that although the magnitude of the drag coefficients is reduced, the discrepancy between the model and observed spectra calculations in the rear quadrants of the hurricane remains the same.

It is well established that the intensity of a tropical cyclone over an open ocean may be significantly affected by the cooling of sea surface temperature caused by air-sea interaction since the tropical cyclones are driven by enthalpy fluxes from the sea and limited mostly by surface drag [78]. While the strong wind of the hurricane leads to evaporation of warm water from the ocean surface that fuels the storm through condensation and latent heat release, as the storm continues to intensify, the increasing wind stress on the ocean's surface generates stronger turbulent mixing that deepens the mixed layer, reduces the sea surface temperature, and causes a reduction of sea surface heat and moisture flux and in turn decrease the intensity of the storm. Thus, the intensity of a tropical cyclone is highly sensitive to the magnitude and spatial distribution of C_d. The results in the Fan and Rogers [76] study have suggested that while the drag coefficients calculated using the wave spectra produced by WAVEWATCH III are comparable to the observations in some quadrants of hurricane, large discrepancies from the observations are found in other quadrants. Thus, the current wave model is not ready for the task of providing accurate drag calculations in the coupled forecast models.

4. Conclusions

This chapter has reviewed the progress in third generation wave models and their applications in tropical cyclone generated surface wave predictions. While we have demonstrated the significant improvements in third generation wave modeling under tropical cyclone condition during the past two decades with increasing accuracy in the model predicted integral properties of the sea (significant wave height, period and direction), the shape of their simulated wave spectra is far less impressive with the modeled wind sea part more problematic than the swells. Although the accuracy of the meteorological forcing is one important factor to blame for these model bias, the substantial degree of empiricism in our wave models based on physical assumptions in our wave theories are also accountable for these model deficiencies, especially for extreme weather conditions such as tropical cyclones.

The dissipation of wind waves in deep water is by far the source term we understand the least. There is hardly any agreement neither on the basic physics of the process nor on the best way, although empirical, to model it. Even though the experimental results exhibit some common features, they are often in serious disagreement with each other and thus does not provide much help in modeling this flux. Thus, given the limited level of knowledge we have on spectral dissipation, it has been used as the tuning knob in the numerical wave models through fittings with the observed wave integral properties (significant wave height, period, direction). However, such an approach of bending the model solutions to match observations may lead to unwanted diverge from the truth we wish to predict. The ultimate solution to this problem still lies within the fundamental improvements in the physical representations of this process.

The coupled atmosphere-ocean-wave models have not shown any convincing improvements in tropical cyclone forecasts, most likely due to the model bias in wave spectra simulations. This does not mean that we should refrain ourselves from this direction. The atmosphere interacts with the ocean through the surface gravity waves. This is how the nature works. Although there will be a long way to go before we can fully understand the dynamical processes that allow us to adequately simulate the wave spectra in the extreme wind regime, this is the right path to take if we aim at a better understanding and modeling of these extreme weather events.

Acknowledgements

This work was funded by the Office of Naval Research, United States of America under program element 0601153N. This article is a contribution of NRL/JA/7320-19-4508 and has been approved for public release. We would like to express our appreciation to the anonymous reviewers for their constructive comments.

Author details

Yalin Fan[1]*, Paul Hwang[2] and John Yu[3]

1 Ocean Science Division, Naval Research Laboratory, Stennis Space Center, MS, United States

2 Remote Sensing Division, Naval Research Laboratory, Washington, DC, United States

3 College of Engineering, University of Michigan, Ann Arbor, MI, United States

*Address all correspondence to: yalin.fan@nrlssc.navy.mil

References

[1] Tolman HL. Validation of a new global wave forecast system at NCEP. In: Edge BL, Helmsley JM, editors. Ocean Wave Measurements and Analysis. Virginia Beach, VA, USA: ASCE; 1998. pp. 777-786

[2] Hasselmann S et al. The WAM model – A third-generation ocean wave prediction model. Journal of Physical Oceanography. 1988;**18**:1775-1810. DOI: 10.1175/1520-0485(1998)018

[3] Booij N, Ris RC, Holthuijsen LH. A third-generation wave model for coastal regions, part 1: Model description and validation. Journal of Geophysical Research. 1999;**104**(C4):7649-7666

[4] Donelan MA, Curcic M, Chen SS, Magnusson AK. Modeling waves and wind stress. Journal of Geophysical Research. 2012;**117**:C00J23. DOI: 10.1029/2011JC00778

[5] Powell MD, Vickery PJ, Reinhold TA. Reduced drag coefficient for high wind speeds in tropical cyclones. Nature. 2003;**422**:279-283

[6] Donelan MA, Haus BK, Reul N, Plant WJ, Stiassnie M, Graber HC, et al. On the limiting aerodynamic roughness of the ocean in very strong winds. Geophysical Research Letters. 2004;**31**:L18306. DOI: 10.1029/2004/GL019460

[7] Makin VK. A note on the drag of the sea surface at hurricane winds. Boundary-Layer Meteorology. 2005;**115**:169-176

[8] Andreas EL. Spray stress revisited. Journal of Physical Oceanography. 2004;**34**:1429-1440

[9] Makin VK, Kudryavtsev VN. Coupled sea surface atmosphere model, 1. Wind over wave coupling. Journal of Geophysical Research. 1999;**104**(C4):7613-7623

[10] Chen SS, Zhao W, Donelan MA, Price JF, Walsh EJ. The CBLAST-hurricane program and the next-generation fully coupled atmosphere-wave-ocean models for hurricane research and prediction. Bulletin of the American Meteorological Society. 2007;**88**:311-317

[11] Fan Y, Ginis I, Hara T, Wright CW, Walsh EJ. Numerical simulations and observations of surface wave fields under an extreme tropical cyclone. Journal of Physical Oceanography. 2009a;**39**:2097-2116

[12] Liu B, Liu H, Xie L, Guan C, Zhao D. A coupled atmosphere-wave-ocean modeling system: Simulation of the intensity of an idealized tropical cyclone. Monthly Weather Review. 2011;**139**:132-152. DOI: 10.1175/2010MWR3396.1

[13] Chen SS, Zhao W, Donelan MA, Tolman HL. Directional wind–wave coupling in fully coupled atmosphere–wave–ocean models: Results from CBLAST-hurricane. Journal of the Atmospheric Sciences. 2013;**70**:3198-3215. DOI: 10.1175/JAS-D-12-0157.1

[14] Hubert WE. Hurricane trajectory forecasts from a non-divergent, non-geostrophic, barotropic model. Monthly Weather Review. 1957;**85**:83-87

[15] Pierson W. Wind generated waves. Advances in Geophysics. 1955;**2**:93-178

[16] Hasselmann S et al. Measurements of wind-wave growth and swell decay during the joint North Sea wave project (JONSWAP). Deutsche Hydrographische Zeitschrift. 1973;**A8**:1-95

[17] WAMDI Group. The WAM model—A third generation ocean wave prediction model. Journal of Physical Oceanography. 1988;**18**:1775-1810

[18] Komen GJ, Cavaleri L, Donelan M, Hasselmann K, Hasselmann S, Janssen P. Dynamics and Modelling of Ocean Waves. Cambridge CB2 2RU, UK: Cambridge University Press; 1994. p. 554

[19] Dell'Osso L, Bertotti L, Cavaleri L. The Gorbush storm in the Mediterranean Sea: Atmospheric and wave simulation. Monthly Weather Review. 1992;**120**:77-90

[20] Bauer E, Hasselmann S, Hasselmann K, Graber HC. Validation and assimilation of Seaseat altimeter wave heights using the WAM wave model. Journal of Geophysical Research. 1992;**97**(C8):12671-12682

[21] Monaldo FM, Beal RC. Comparison of SIR-C SAR wavenumber spectra with WAM model predictions. Journal of Geophysical Research. 1998;**103**(C9):18815-18825

[22] Tolman HL. Alleviating the garden sprinkler effect in wind wave models. Ocean Model. 2002;**4**:269-289

[23] Tolman HL. User manual and system documentation of WAVE-WATCH-III version 1.18. Tech. Note 166, 1999; Ocean Modeling Branch, NCEP, National Weather Service, NOAA, U.S. Department of Commerce. p. 110. Available from: http://polar.wwb.noaa.gov/waves/wavewatch

[24] Tolman HL. Validation of WAVEWATCH III version 1.15 for a global domain. NOAA/NWS/NCEP/OMB Tech. Note 213. 2002a. p. 33. Available from: http://polar.ncep.noaa.gov/waves/references.html

[25] Tolman HL, Balasubramaniyan B, Burroughs LD, Chalikov D, Chao YY, Chen HS, et al. Development and implementation of wind-generated ocean surface wave models at NCEP. Weather and Forecasting. 2002;**17**:311-333

[26] Wingeart KM, O'Reilly WC, Herbers THC, Wittmann PA, Jenssen RE, Tolman HL. Validation of operational global wave prediction models with spectral buoy data. In: Edge BL, Hemsley JM, editors. Ocean Wave Measurement and Analysis. San-Francisco, CA, USA: ASCE; 2001. pp. 590-599

[27] Booij N, Holthuijsen LH, Ris RC. The "SWAN" wave model for shallow water. In: Proceedings of 25th Conference on Coastal Engineering, 2-6 September 1996. Orlando, Florida; 1996. DOI: 10.1061/9780784402429.053

[28] Holthuijsen LH, Booij N, Padilla-Hernandez R. A curvilinear, third-generation coastal wave model. In: Proceedings of the Coastal Dynamics '97 Conference, 8 June 1997. Plymouth, United Kingdom: American Society of Civil Engineers; 1998. pp. 128-136

[29] Chen SS, Curcic M. Ocean surface waves in hurricane Ike (2008) and super- storm Sandy (2012): Coupled model predictions and observations. Ocean Modelling. 2015;**103**:161-176. DOI: 10.1016/j.ocemod.2015.08.005

[30] Liu Q, Babanin A, Fan Y, Zieger S, Guan C, Moon I. Numerical simulations of ocean surface waves under hurricane conditions: Assessment of existing model performance. Ocean Model. 2017;**118**:73-93. DOI: 10.1016/j.ocemod.2017.08.005

[31] Ou S-H, Liau J-M, Hsu T-W, Tzang S-Y. Simulating typhoon waves by SWAN wave model in costal waters of Taiwan. Ocean Engineering. 2002;**29**:947-971

[32] Wyatt LR. The effect of fetch on the directional spectrum of Celtic Sea storm waves. Journal of Physical Oceanography. 1995;**25**:1550-1559

[33] Holt B, Liu AK, Wang DW, Gnanadesikan A, Chen HS. Tracking

storm-generated waves in the Northeast Pacific Ocean with ERS-1 synthetic aperture radar imagery and buoys. Journal of Geophysical Research. 1998;**103**(c4):7917-7929

[34] Wright CW et al. Hurricane directional wave spectrum spatial variation in the open ocean. Journal of Physical Oceanography. 2001;**31**:2472-2488

[35] Walsh EJ et al. Hurricane directional wave spectrum spatial variation at landfall. Journal of Physical Oceanography. 2002;**32**:1667-1684

[36] Moon I-J, Ginis I, Hara T, Tolman H, Wright CW, Walsh EJ. Numerical simulation of sea-surface directional wave spectra under hurricane wind forcing. Journal of Physical Oceanography. 2003;**33**:1680-1706

[37] Chao YY, Alves J-HGM, Tolman HL. An operational system for predicting hurricane-generated wind waves in the North Atlantic Ocean. Weather and Forecasting. 2005;**20**:652-671

[38] Tolman HL, Alves J-HGM. Numerical modeling of wind waves generated by tropical cyclones using moving grids. Ocean Modelling. 2005;**9**:305-323

[39] Tolman HL, Alves JHGM, Chao YY. Operational forecasting of wind-generated waves by hurricane Isabel at NCEP. Weather and Forecasting. 2005;**20**:544-557

[40] Black PG et al. Air–sea exchange in hurricanes: Synthesis of observations from the coupled boundary layer air–sea transfer experiment. Bulletin of the American Meteorological Society. 2007;**88**:357-374

[41] Powell MD, Houston SH, Amat LR, Morisseau-Leroy N. The HRD real-time hurricane wind analysis system. Journal of Wind Engineering and Industrial Aerodynamics. 1998;**77-78**:53-64

[42] Fan Y, Ginis I, Hara T. The effect of wind-wave-current interaction on air-sea momentum fluxes and ocean response in hurricanes. Journal of Physical Oceanography. 2009b;**39**(4):1019-1034

[43] Montoya RD, Osorio AA, Ortiz Royero JC, Ocampo-Torres FJ. A wave parameters and directional spectrum analysis for extreme winds. Ocean Engineering. 2013;**67**:100-118

[44] Ortiz JC, Mercado A. An intercomparison of SWAN and Wavewatch III models with data from NDBC-NOAA buoys at oceanic scales. Coastal Engineering Journal. 2008;**50**(1):47-73

[45] Snyder RL, Dobson FW, Elliott JA, Long RB. Array measurements of atmospheric pressure fluctuations above surface gravity waves. Journal of Fluid Mechanics. 1981;**102**:1-59

[46] Komen GJ, Hasselmann S, Hasselmann K. On the existence of a fully developed wind-sea spectrum. Journal of Physical Oceanography. 1984;**14**:1271-1285

[47] Tolman HL, Chalikov DV. Source terms in a third-generation wind-wave model. Journal of Physical Oceanography. 1996;**26**:2497-2518

[48] Tolman HL. Distributed memory concepts in the wave mode WAVEWATCH III. Parallel Computing. 2002b;**28**:35-52

[49] Chalikov DV, Belevich MY. One-dimensional theory of the wave boundary layer. Boundary-Layer Meteorology. 1993;**63**:65-96

[50] Phillips OM. The equilibrium range in the spectrum of wind-generated

waves. Journal of Fluid Mechanics. 1958;**4**(4):426-434

[51] Bidlot J-R. Present status of wave forecasting at E.C.M.W.F. In: Proceedings of ECMWF Workshop on Ocean Wave Forecasting. U. K.: Reading; 25-27 June 2012

[52] Miles JW. On the generation of surface waves by shear flows. Journal of Fluid Mechanics. 1957;**3**(2):185-204

[53] Janssen PAEM. Quasi-linear theory of wind-wave generation applied to wave forecasting. Journal of Physical Oceanography. 1991;**21**:1631-1642

[54] Janssen P. The Interaction of Ocean Waves and Wind. Cambridge CB2 2RU, UK: Cambridge University Press; 2004. p. 300

[55] Hasselmann K. On the spectral dissipation of ocean waves due to whitecapping. Boundary-Layer Meteorology. 1974;**6**(1-2):107-127

[56] Phillips OM. Spectral and statistical properties of the equilibrium range in wind-generated gravity waves. Journal of Fluid Mechanics. 1985;**156**:505-531

[57] Jenkins AD. A Lagrangian model for wind- and wave-induced flux of near-surface currents. Coastal Engineering. 1987;**11**:513-526

[58] Banner ML, Young IR. Modeling spectral dissipation in the evolution of wind waves. Part I: Assessment of existing model performance. Journal of Physical Oceanography. 1994;**24**:1550-1571

[59] Banner ML, Babanin AV, Young IR. Breaking probability for dominant waves on the sea surface. Journal of Physical Oceanography. 2000;**30**:3145-3160

[60] Banner ML, Gemmrich JR, Farmer DM. Multiscale measurements of ocean wave breaking probability. Journal of Physical Oceanography. 2002;**32**:3364-3375

[61] Babanin AV, Young IR, Banner ML. Breaking probabilities for dominant surface waves on water of finite constant depth. Journal of Geophysical Research. 2001;**106**(C6):11659-11676

[62] Babanin AV, Chalikov D, Young IR, Savelyev I. Predicting the breaking onset of surface water waves. Geophysical Research Letters. 2007;**34**:L07605. DOI: 10.1029/2006GL029135

[63] Ardhuin F, Rogers WE, Babanin AV, Filipot J, Magne R, Roland A, et al. Semiempirical dissipation source functions for ocean waves. Part I: Definition, calibration, and validation. Journal of Physical Oceanography. 2010;**40**:1917-1941

[64] Babanin AV. Breaking and Dissipation of Ocean Surface Waves. Cambridge CB2 2RU, UK: Cambridge University Press; 2011. p. 480

[65] Zieger S, Babanin AV, Rogers WE, Young IR. Observation based source terms in the third-generation wave model WAVEWATCH. Ocean Modelling. 2015;**96**:2-25

[66] Hasselmann S, Hasselmann K, Allender JH, Barnett TP. Computations and parameterizations of the nonlinear energy transfer in a gravity-wave spectrum. Part II: Parameterizations of the nonlinear energy transfer for application in wave models. Journal of Physical Oceanography. 1985;**15**(11):1378-1392

[67] van Vledder GP. The WRT method for the computation of non-linear four-wave interactions in discrete spectral wave models. Coastal Engineering. 2006;**53**(2-3):223-242. DOI: 10.1016/j.coastaleng.2005.10.011

[68] Tolman HL. A generalized multiple discrete interaction approximation

for resonant four-wave interactins in wind wave models. Ocean Modelling. 2013;**70**:11-24. DOI: 10.1016/j.ocemod.2013.02.005

[69] Emanuel K. Tropical cyclones. Annual Review of Earth and Planetary Sciences. 2003;**31**:75-104. DOI: 10.1146/annurev.earth.31.100901.141259

[70] Zambon JB, He R, Warner JC. Investigation of hurricane Ivan using the coupled ocean-atmosphere-wave-sediment transport (COAWST) model. Ocean Dynamics. 2014;**64**:1535-1554. DOI: 10.1007/s10236-014-0777-7

[71] Varlas G, Katsafado P, Papadopoulos A, Korres G. Implementation of a two-way coupled atmosphere-ocean wave modeling system for assessing air-sea interaction over the Mediterranean Sea. Atmospheric Research. 2018;**208**:201-217. DOI: 10.1016/j.atmosres.2017.08.019

[72] Gopalakrishman S, et al. 2016 HFIP R&D Activities Summary: Recent Results and Operational Implementation. NOAA Tech. Rep. HFIP2017-1; 2017. p. 61. Available from: http://www.hfip.org/documents/HFIP_AnnualReport_FY2016.pdf

[73] Gopalakrishnan S, et al. 2017 HFIP R&D activities summary: Recent results and operational implementation. NOAA Tech. Rep. HFIP2018-1; 2018. p. 41. Available from: http://www.hfip.org/documents/HFIP_AnnualReport_FY2017.pdf

[74] Gopalakrishnan S, et al. 2018 HFIP R&D activities summary: Recent results and operational implementation. NOAA Tech. Rep. HFIP2018-1; 2019. p. 47. Available from: http://www.hfip.org/documents/HFIP_AnnualReport_FY2018.pdf

[75] Gopalakrishnan S, et al. 2019 HFIP R&D activities summary: Recent results and operational implementation. NOAA Tech. Rep. HFIP2018-1; 2020. p. 42. Available from: http://www.hfip.org/documents/HFIP_AnnualReport_FY2019.pdf

[76] Fan Y, Rogers WE. Drag coefficient comparisons between observed and model simulated directional wave spectra under hurricane conditions. Ocean Modelling. 2016;**102**:1-13. DOI: 10.1016/j.ocemod.2016.04.004

[77] Donelan MA, Babanin AV, Young IR, Banner ML. Wave-follower field measurements of the wind-input spectral function. Part II: parameterization of the wind input. Journal of Physical Oceanography. 2006;**36**:1672-1688

[78] Ginis I. Tropical cyclone-ocean interactions. In: Perrie W, editor. Atmosphere-Ocean Interactions. 1st ed. Ashurst, Southampton SO40 7AA, UK: WIT Press; 2002. pp. 83-114

Chapter 7

Simplified Methods for Storm Surge Forecast and Hindcast in Semi-Enclosed Basins: A Review

Davide Pasquali

Abstract

It is widely known that small and semi-enclosed basins could be inclined to storm surge events. This is mainly due to either the meteorological exposition, to the presence of a continental shelf or to their shape. These storm surges can induce coastal flooding and consequent problems in terms of infrastructure stability and damage to touristic activities or, in some cases, threaten human life. Therefore, in order to manage the risk, coastal managers or policymakers need to have forecast or hindcast tools. They must help to take preventive actions that may be done previously to the occurrence of natural phenomena and to carry out simultaneous actions useful during the occurrence of the event. This work aims at answering these necessities presenting a review of two methods for storm surge forecast and hindcast in semi-enclosed basins.

Keywords: storm surge forecasting, storm surge hindcasting, sea level variations, artificial neural networks, extreme events, generalized Pareto distribution, coastal flood risk, risk assessment, sea level rise, climate change

1. Introduction

The last few years have seen the increase of human settlement in the coastal areas, for social, touristic or economic reasons. This phenomenon leads to a consequent increase in human occupation of coastal areas (e.g. [1]). All these factors, combined with the attention paid to the sea level rise scenarios (e.g. [2]) and with the problem of coastal erosion (e.g. [3–5]), justify the increasing attention paid to storm surge events and related coastal flooding.

As it is well known, the tide level is the superposition of a harmonic component related to the mutual influence of the earth, sun and moon and a meteorological one. The first is purely deterministic and has been studied since the last years of the nineteenth century. Indeed, since the observation of the phenomenon started before Christ, only after Newton's theory [6] researchers as George Darwin [7] provided the first mathematical description of the tides. However, the first model of the tides is the "harmonic theory" developed in 1927 by Doodson [8]. It considers the tides as a superposition of sinusoidal constituents, whose frequencies are referred to as those evaluated on the basis of astronomical forces. The amplitude, instead, is influenced by the shallow water conditions, the coastal effects and morphological phenomena due to the interaction between waves and bottom [9].

For a more comprehensive description of the "harmonic theory", the reader may refer to Godin, McCully, and Pugh [9–11].

The difference between the deterministic component and the total measured level oscillation can be related to meteorological phenomena and may be defined as *storm surge*. It is also referred to as *meteorological component* or *residual*. In European literature, the term storm surge is commonly used [12].

Generally speaking, the generation of storm surges is (mainly) due to pressure gradients and wind set-up. The effect of the wind is to push the water in the principal direction of the wind causing an increase in sea level. The other factor that induces variation in sea level is the barotropic field (e.g. pressure anomalies) causing the physical phenomenon of the "inverse barometric effect" that is the increase of mean sea level as the pressure decreases (e.g. 1 cm for each hPa).

Talking about small and semi-enclosed basins (e.g. Adriatic Sea, Black Sea, Caspian Sea, Great Lakes, etc.), there is another effect contributing to the level increase. This effect may be attributed to the case in which meteorological perturbations persist for a long time over the basin (e.g. until several days). In those situations, there are two main consequences: The first is that it could be difficult to forecast the storm surge (e.g. [13]), and the second is related to the dynamic of the basin. Indeed, if the basin is semi-enclosed, its natural modes, i.e. seiches, may be excited, and level oscillations may persist for several days in the whole basin area (e.g. [14–16]).

While the astronomical component can be estimated and reconstructed by means of standard techniques (based on the theory of harmonic analysis) (e.g. [11, 17, 18]) performed by using measured level time series, the reconstruction of storm surge events, with both forecast and hindcast purposes, is not deterministic and requires more effort in its evaluation.

The topic of storm surge and their forecast has been investigated in the past by many researchers (e.g. [19–22]), and the importance of the topic is also highlighted observing that there are countries that are being equipped with early warning systems (e.g. [23]).

From a practical point of view, there are three ways to study storm surges: pure numerical approaches (i.e. circulation models), statistical approaches (i.e. artificial neural networks) and mixed approaches.

In the case of a pure numerical approach, the aim is to focus the attention on the ability of the model to reproduce the physical processes (e.g. [24]). These methods are physics-based and are often referred to as "dynamical method" (e.g. [25]). They numerically solve the classical mathematical set of equations composed by the continuity equation and the equation of motion where the initial and boundary conditions are given by a meteorological model (e.g. [25]). As pointed out by Vilibić et al. (e.g. [25]), the first examples of dynamical methods, at least for the Adriatic Sea, are [26, 27], for the only Adriatic area and for the entire Mediterranean Sea, respectively. An improvement in terms of forecast reliability (also in terms of forecast window) is the use of ensemble-based prediction systems (e.g. [28]) that allow having a more consistent forecast than that obtained with a single deterministic one.

Statistical approaches, instead, are based on the use of regression models to estimate a series of predictors and weights to forecast the desired variable (i.e. water level). The database is usually composed by observed or forecast and/or hindcast data given by a meteorological model (e.g. [29–33]).

Numerical models are accurate, with great reliability, as they are physics-based. However they involve high computational costs compared to the statistical ones. These are fast, show an acceptable reliability, but have no physics inside.

A possible way to overcome these problems is the use of the last category of approaches, the "mixed" ones. They statistically correct the results coming from numerical models with the aim to reduce the computational costs. The idea is to use numerical models accepting low reliability using meshes with lower resolution (i.e. lower computational costs) and correcting the obtained results by means of statistical tools (e.g. [34–36]).

In all the three cases (dynamical, statistical and mixed approaches), meteorological data provided by global general circulation models (e.g. European Centre for Medium-Range Weather Forecasts (ECMWF), the Meteorological Research Institute model (MRI-AGCM3.2) [37, 38]) have to be used, so the final reliability is often related to those of the GCMs.

This chapter aims to propose a review of two simplified methods [36, 39] finalized to forecast and hindcast storm surge levels. Both approaches are mixed, so they have a first physics-based approach where the water level due to the wind is evaluated (once at all) and a separate step in which the obtained results are corrected (i.e. the barotropic effects are considered) by means of statistical techniques. The chapter is structured as follows. Section 2 describes the general outline of the two methods, Section 2.2 illustrates the forecast method, while Section 2.3 details the hindcast one. The applications of the two approaches are described in Section 3. Concluding remarks close the chapter.

2. Methods

2.1 Outline

The whole idea behind the presented forecast and hindcast methods lies in the theory of linear dynamic systems (e.g. [40]). The concept is to consider, at least for technical purposes, the dynamics of semi-enclosed basins (e.g. [41]) as linear. The linearity allows to compute the unit response function of the basin and use it to determine (by using the convolution integral) the response of water level due to any wind time series.

More in detail, at this step a basin is discretized in small areas. In each area, the wind must be considered homogeneous and constant. The total number of the areas must be chosen in order to capture the variation of wind field across the basin (i.e. gradients). Considering a generic point of interest in the domain (hereinafter referred as POI), the elevation $\eta(t)$ due to one wind stress impulse acting on a generic area i of the domain will be

$$\eta_i(t) = UF_i^U(t) + VF_i^V(t) \tag{1}$$

where $F_i^U(t)$ and $F_i^V(t)$ are the unit response functions induced by wind with a duration equal to Δt acting on area i, while U and V are the components of wind stress impulse along with the two Cartesian directions (zonal and meridional, respectively) acting on the same area.

Considering M wind stress impulses, Eq. (1) becomes

$$\eta_i(t) = \sum_{j=1}^{j\le M} U_{ij}F_i^U(t - j\Delta t + \Delta t) + V_{ij}F_i^V(t - j\Delta t + \Delta t) \tag{2}$$

The value obtained with Eq. (2) is the water elevation of a POI due to a series of wind stress impulses U_{ij} and V_{ij} happened between $t = (j-1)\Delta t$ and $t = j\Delta t$ on area i. The level due to the contribution of N areas can be easily calculated by using the superposition of the role of each area; therefore, Eq. (2) must be modified to as

$$\eta(t_k) = \sum_{i}^{N} \sum_{j=1}^{j \leq M} U_{ij}^{(\tau)} F_i^U (t_k - j\Delta t + \Delta t) + V_{ij}^{(\tau)} F_i^V (t_k - j\Delta t + \Delta t) \quad (3)$$

with $t_k = (k-1)\Delta\tau$.

The values of U and V can be forecast or hindcast data provided by global general circulation models and must be taken at a point at each area.

The unit response functions F_i^U and F_i^V, instead, can be evaluated using a numerical model (e.g. [42, 43]). It has to be stressed that the computation of the unit response functions has to be done once for all for each considered basin. In this way, it is possible to limit the computational costs of the methods.

However, the computed level is due to the wind field effect and does not consider the role of the barotropic field in the storm surge generation. For this reason, it can be viewed as a "raw level".

The pressure field is considered using statistical techniques. The forecast and the hindcast models have the physics-based module as a common part but correct the raw level in a different manner.

2.2 Forecast method statistical correction

As previously underlined the statistical corrections are often carried out using regression models or, alternatively, artificial neural networks (ANNs).

In the proposed method, the use of a series of ANNs aimed at correcting the forecast for each lead time is suggested. Without claiming to be exhaustive, the ANNs could be defined as a statistical tool that reproduces the human ability of learning. They are made up of a layer of input neurons that, interacting with connections characterized by their own weight (i.e. hidden layers, activation function, etc.), produce one or more output neurons.

The learning phase is basically the way in which the system acquires information in order to predict future events on the basis of past experience. Mathematically, this phase consists of a training aimed at reducing the mean square error (MSE) between the output neuron(s) and the wished output by changing the weights of the links between neurons. The methodology is iterative, and it can be stopped only (after a fixed number of training epochs) when the MSE is lower than a given threshold. After the learning phase, the net needs to be tested in order to check its performance. If results are not consistent with the desired accuracy, it is possible to repeat the training phase. It has to be noticed that this phase is a "black box", so results from a training and another one could be very different from each other. At the end of this process, the network can be used in operational situations.

In the presented case, the choice was to use a series on ANNs instead of only one because each ANN operates only on one lead time (to correct a forecast of 48 h, 48 ANNs are needed).

A crucial point in ANNS is the choice of input neurons. In the case at hand, the use of (a) the raw level time series, (b) recent level measurements at the POI and (c) the pressures at the centre of each area used within the frame of the physics-based module is proposed. Of course, a sensitivity analysis could be useful in order to perform this choice in other cases; i.e. the procedure is site-dependent.

2.3 Hindcast method statistical correction

The statistical correction in the hindcast method is different from that of the forecast one. The reason lies in the different purposes of the method. Indeed, when the measured data time series are not long enough to be considered representative (i.e. statistically) for high return periods, the use of hindcast methods is the only way to work with a subset of reliable data.

In these cases, the interest is not focused on the timing of the reconstructed time series, but on the reliability in reproducing the extreme events.

For these reasons, the correction of the raw level (obtained using reanalysis data) is carried out by adding the pressure field using a coefficient C_p that can be estimated by comparing synchronous observed residual levels and pressure values at a generic point of interest. This means that Eq. (3) is modified as follows:

$$\eta(t_k) = \sum_{i}^{N} \sum_{j=1}^{j \leq M} U_{ij}^{(\tau)} F_i^U(t_k - j\Delta t + \Delta t) + V_{ij}^{(\tau)} F_i^V(t_k - j\Delta t + \Delta t) + C_p \Delta p(t_k) \quad (4)$$

where $\Delta p(t_k)$ represents the atmospheric pressure anomaly.

Due to the aim of the method (i.e. achieve reliable estimates of extreme events), a correction on the maximum hindcast error for a given temporal window has to be defined. Having available measured data, it is possible to define the maximum error (ϵ_{max}) by comparing the maximum (not necessary synchronous) measured (η_{max}^M) and hindcast values (η_{max}^H) occurring within a time frame of a given duration. Following (e.g. [44]) this error could be defined as the relation between η_{max}^M and η_{max}^H, and this allows to define a calibration coefficient as

$$C_{cal} = \frac{\eta_{max}^M}{\eta_{max}^H}. \quad (5)$$

Following the technique proposed by [44], it is possible to evaluate the quantile of the ECDF of the random variable C_{cal}. This quantile should be viewed as the comparison between extreme values of observation and the corrected hindcast series. It is clear that this method requires the availability (even for a few years) of measured tide levels and pressure values.

3. Application of the methods

This section aims to illustrate the application of the described methods showing their general performances by means of two applications to real cases. The forecast method is applied to the northern part of the Adriatic Sea, while the hindcast one to the southern Adriatic Italian coast.

3.1 Forecast method: the case of Venice

This section shows an example of the application of the forecast method to a POI located in the northern part of the Adriatic Sea. For this basin, storm surges are mainly due to Atlantic perturbations (i.e. cyclones). Due to the presence of the continental shelf and considering thermal effects, the perturbations are amplified [45]. A famous example of these effects is Venice and the phenomenon of "acqua alta" causing the partial or total flooding of the city with damage to historical monuments, economy and private buildings. The recent flooding event of November 2019 has been supposed to damage the city for 4 billion of euros.

The physical characteristics of the weather conditions can induce resonance phenomena (e.g. [14–16]) with level oscillations persisting for several days in the whole basin [46].

The worst situation for the Adriatic Sea in terms of storm surge is when the perturbations come from south-east (i.e. "sirocco winds") and propagate along the main axis of the basin (e.g. [45]).

The Adriatic Sea has been discretized in 19 areas ($N = 19$). The forecast wind and pressure data have been taken by the European Centre of Medium-Range Weather Forecast. The dimension of each area is 0.25° (see **Figure 1**).

The unit wind response functions have been estimated for each of the 19 areas using the "regional ocean modelling system" (ROMS, e.g. [42, 47]). Due to the particular geographical conditions and to the prevalent wind directions (i.e. Sirocco), only the wind stress component acting along the main axis of the Adriatic Sea (i.e. $\simeq 324°$N) has been used. A sensitivity analysis shows that results are the same using both U and V wind components but limiting the computational effort. This means that also the constitutive equation (3) reads as

$$\eta(t_k) = \sum_{i}^{N} \sum_{j=1}^{j \le M} R_j F_i(t_k - j\Delta t + \Delta t) \tag{6}$$

where R_j is the projection of the wind stress vector along the main axis of the basin and F_i is the unit response function for unit wind stress blowing along the main axis of the basin at the ith area.

About the simulations, the grid resolution is 3′ (i.e. about 5500 m, 175 × 185 computational points) including all the Adriatic Sea and a portion of the Ionian Sea (**Figure 1**). The "Etopo1" bathymetry has been used [48]. The coasts have been modelled with wall boundary conditions, while in the southern part, a radiation condition has been imposed (e.g. [46]). Each of the 19 simulations differs from others for the area in which the wind has been imposed. The duration of the wind

Figure 1.
Illustration of the discretization of the basin. The black circles indicate the POIs, while the grey arrow indicates the direction of the main axis of the basin.

impulses is 6 hours (ECMWF resolution), and the response functions were given with a time resolution equal to $\Delta\tau$ = 900 s. **Figure 2** shows an example of the computed response functions for two areas located in two different locations in the basin.

The wind stress values have been computed on the basis of wind speed by using the relationship proposed by Drago and Iovenitti [49]

$$R_j = \gamma_s W_l |W| \quad (7)$$

where W_l is the wind speed component along the main direction $\simeq$324°N and W is the actual wind speed. The factor γ_s is linked to the wind speed (e.g. [50]):

$$\gamma_s = 6.9 \cdot 10^{-4} + 7.5 \cdot 10^{-5} |W| \quad (8)$$

Figure 3 shows the performances of the forecast method. The figure shows some typical results illustrating the forecasted storm surge level (dashed grey line), the

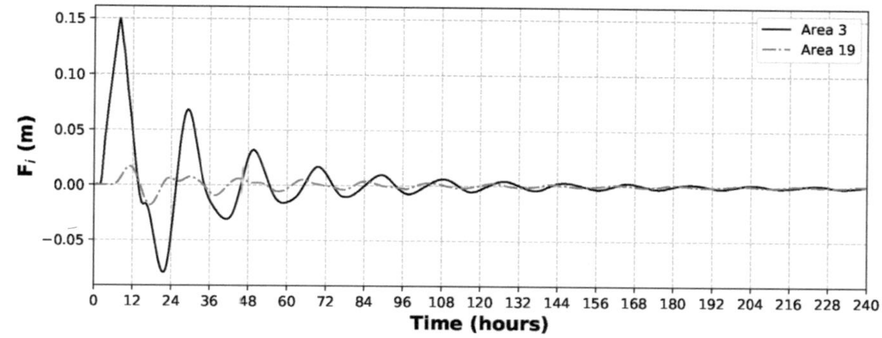

Figure 2.
Example of the computed response functions for two different areas.

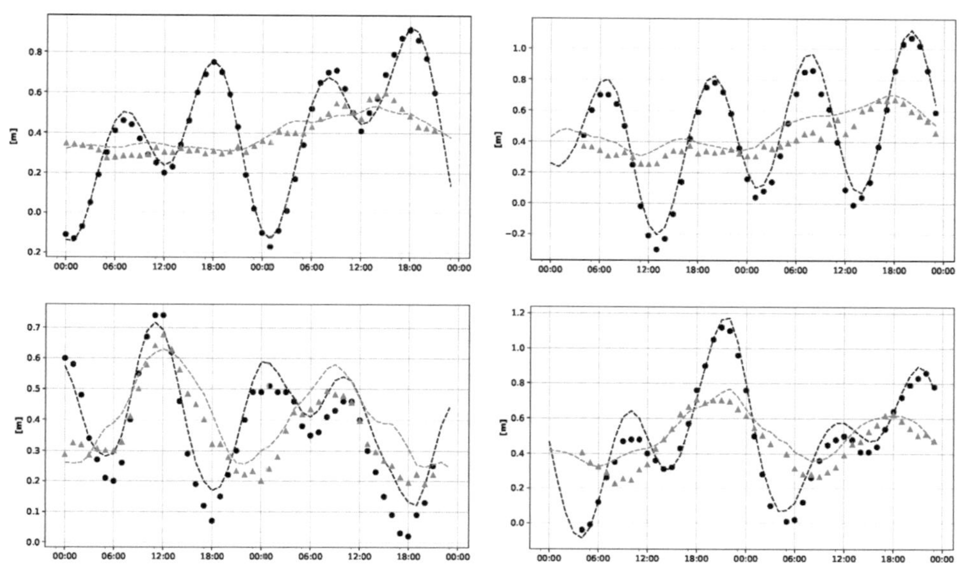

Figure 3.
Comparison between the forecasted storm surge level (dashed grey line), the measured storm surge level (triangle symbols), the forecasted total tide level (dashed black line) and the measured total tide level (black circles).

measured storm surge level (triangle symbols), the forecasted total tide level (dashed black line) and the measured total tide level (black circles). The measurements have been taken using the records in the mareographic station (45.41°N, 12.44°E) located on Lido mouth (Italian National Mareographic Network). It is also illustrated in the figure the total measured tide level (grey cross symbols) and the forecasted one (solid grey lines). The harmonic component has been evaluated by means of [17] considering seven components (M2, S2, N2, K2, K1, O1, P1) as suggested in the literature (e.g. [51]).

The statistical correction have been made training 48 ANNs, one for each increasing lead time ($\Delta t_n = n$-hours, $n = 1, 2, \ldots, 48$). The used ANNs are multilayer networks with an input vector, two hidden layers and one output value (the storm surge). The training has been made using a back-propagation algorithm. In this case, a Levenberg–Marquardt algorithm has been used. The learning phase of the ANNs covered 3 years (2009–2011); the testing period is referred to as the year 2012, while the validation period is the year 2013.

As it is possible to see, inspecting **Figure 3**, the training phase confirms the good performances of the algorithm, and, in the validation step, at least from a qualitative point of view, it is possible to appreciate the accuracy of the model.

Talking about quantitative performances, a series of statistical parameters have been evaluated. Mean (μ) and standard deviation (σ) of the differences between predicted and observed total tide have been computed for the years 2009–2011 (training period), year 2012 (testing period) and year 2013 (validation period, e.g. [34, 47]). **Table 1** summarizes the results of the statistical analysis. It is important to observe that the absolute value of the mean is always lower than about 0.04 m, while the standard deviation (that increases as the lead time increases) ranges between 0.05 m and 0.10 m.

The comparison between the obtained results and those available in the literature (e.g. [34]) reveals that the gained reliability is satisfactory if the simplicity and computational costs of the method are considered.

Lead time [hours]	μ [m]	σ [m]	Lead time [hours]	μ [m]	σ [m]
1	−0.008	0.045	13	0.008	0.077
2	−0.004	0.054	14	0.013	0.083
3	0.017	0.072	15	0.025	0.091
4	0.012	0.077	16	0.028	0.095
5	0.010	0.077	17	0.043	0.097
6	0.020	0.069	18	0.012	0.079
7	0.014	0.062	19	−0.002	0.070
8	0.019	0.063	20	0.006	0.064
9	0.025	0.062	21	0.019	0.073
10	0.011	0.062	22	0.026	0.082
11	0.000	0.062	23	0.001	0.077
12	0.000	0.066	24	−0.006	0.078

Table 1.
Mean (μ) and standard deviation (σ) of the differences between foreseen and measured total tide level as a function of lead time.

3.2 Hindcast method: the case of Manfredonia

This section aims at showing an example of the application of the hindcast method to a POI located in the south of Italy, in the Manfredonia Gulf (41.38°N, 15.55°E). When the wind comes from south or south-east rivers, flow is influenced by the downstream boundary condition, and several areas are flooded. These events cause damage to economic activities, private houses, etc. Moreover, the coastal area on the Gulf suffers from erosion. Although standard protection structures (e.g. [52, 53]) have been rolled out, erosion problems are still unresolved.

In this scenario, the utility of having a tool to perform hazard analysis is clear [3].

As observed, this method relies on the same mathematical hypotheses of the forecast one using a different statistical approach to correct the raw data coming from the dynamical step.

The response functions of the basin have been evaluated in the same way as described in Section 3.1 and extracted in a different point of interest (see **Figure 1**).

Following the description in Section 2.3, to apply the hindcast method, a set of data has been known. In the present case, the wind and atmospheric pressure data are referred to the ERA-Interim database (European Centre for Medium-Range Weather Forecasts (e.g. [54])). The spatial resolution of ERA-Interim data is 0.75°, while the response function of the basin, as described above, has another resolution equal to 3′. To overcome this problem, the values of wind and pressure (acting at the centre of each area) have been evaluated performing a linear interpolation.

The tidal data are those collected by means of the tidal gauge station owned by the Apulia Region Meteomarine Network (also referred to as SIMOP, e.g. [55]). It collects wave, wind and tidal data along the Apulian coasts [56]. This station does not gather the measures of atmospheric pressure. Then, in order to compute the term related to pressure gradients, this data have been taken in two locations near Manfredonia: Vieste and Bari (see **Figure 4**) where two tidal gauges of the National Mareographic Network are installed.

Figure 4.
Sketch of the study area. Circles indicate the point of interest of Manfredonia and the locations mentioned in the paper.

As for the case of Venice, the raw level can be evaluated using a simplified version of Eq. (4) considering the projection of the wind stress along the principal orientation of the basin (i.e. $\simeq 324°$N). The modified equation reads as

$$\eta(t_k) = \sum_{i}^{N} \sum_{j=1}^{j \le M} W_{ij} F_i^W (t_k - j\Delta t + \Delta t) + C_p \Delta p(t_k) \qquad (9)$$

where W_{ij} is the projection of the wind stress impulse along the principal orientation of the basin and $\Delta p(t_k)$ is the pressure anomaly.

The coefficient C_p is the correlation factor between the residual levels estimated by means of the dynamic approach and the related measured pressure anomalies (i.e. pressure and total tide level are mandatory). In this case, as previously declared, the pressure measures are referred to as those acquired in the mareographic stations in Bari and Vieste.

Due to the proximity of the stations to the POI, as might be expected, pressure measurements are in agreement. For a more detailed description, the reader may refer to [39]. It is possible to reach the same conclusion considering the quantiles of the measures in Vieste and Bari (see **Figure 5**) and comparing the quantiles of the pressure extracted from the ERA-Interim database at the centre of area 12 (that is the area the POI belongs to, see **Figure 1**) against the quantiles of the pressure observed at Bari (P_{BARI}, left panel) and at Vieste (P_{VIESTE}, right panel). **Figure 6** shows the results of the comparison between the quantiles of the ERA-Interim data and observations.

Based on these outcomes, arguing that the field pressure is almost the same in the area between Bari and Vieste, for Manfredonia, a value of C_p equal to 0.905 (the average value estimated for Bari and Vieste) has been considered.

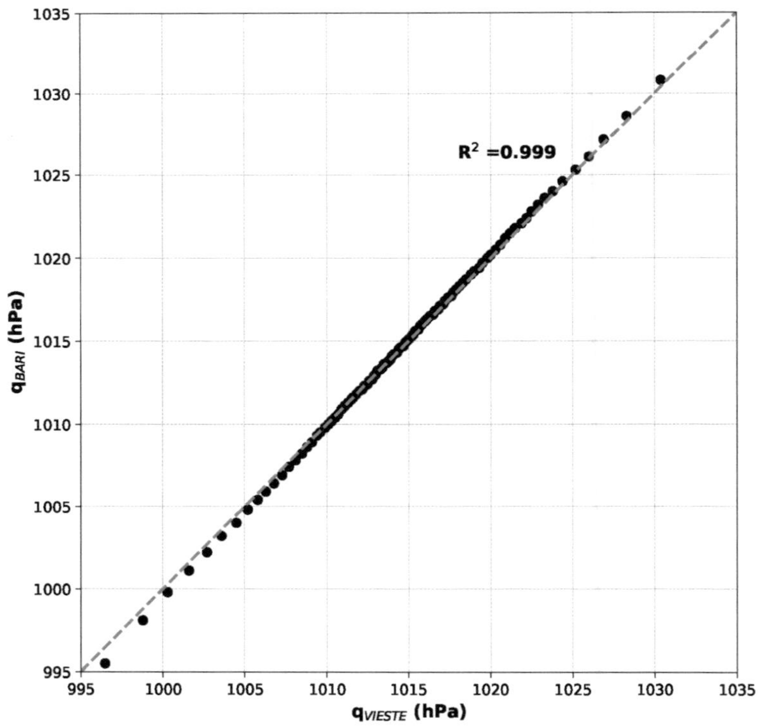

Figure 5.
Quantiles of the measured pressure at Bari (P_{BARI}) and Vieste (P_{VIESTE}). The dashed line refers to the perfect fit line.

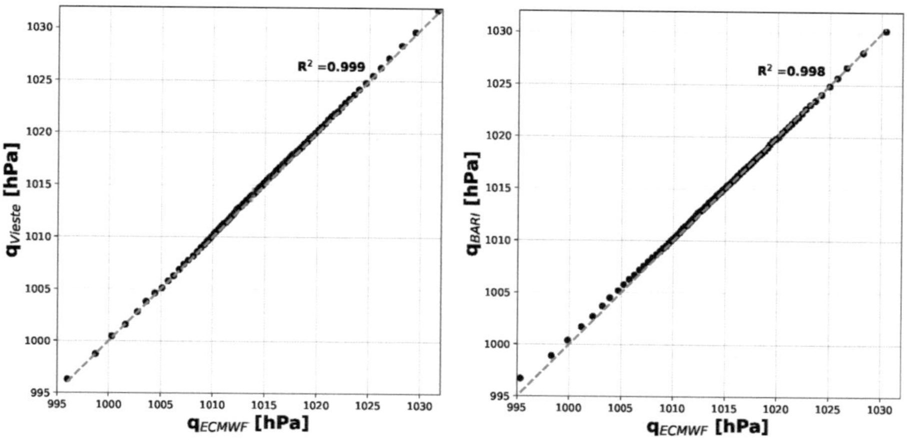

Figure 6.
Comparison between quantiles of the ERA-Interim data and observations at the centre of area 12 against the pressure observed at Bari (P_{BARI}, left panel) and Vieste (P_{VIESTE}, right panel). The perfect fitting line is in marked in grey.

A total of 39 years (from 1979 to 2017) of residual tide levels have been reconstructed by means of Eq. (9). Also in this case, although Eq. (9) considers also the pressure anomalies (i.e. using the term $C_p\Delta p(t_k)$), results strongly depend on the reliability of the selected reanalysis data. In the presented application, ERA-Interim data have been used. As underlined by [57], this database tends to underestimates the hindcast time series. Therefore, also in this case, a statistical correction must be made. Considering that the main aim of this method is to build hindcast time series to be used for return level estimation (i.e. correct hindcast of extreme values, see Section 2.3), the calibration coefficient was evaluated considering the population of the random variable C_{cal} given by Eq. (5). The selection has been made by matching the quantiles of the probability density functions of the hindcast and observed extreme values.

The extreme extraction has been performed by means of a peak over threshold (POT) analysis. The obtained data have been used to define the generalized Pareto distribution (GPD) (e.g. [58]). The threshold selection has been made following the standard technique proposed by [58].

Varying the calibration coefficient C_{cal} ranging from 1.0 up to 2.0, the return levels of the hindcast time series and of the observed values (Xr_M) have been computed. Measured data show a threshold equal to 0.10 m with 162 values exceeding the threshold, while the estimated GPD parameters are $\xi = 0.22$ (shape parameter) and $\sigma = 0.03$ (scale parameter). The calibration coefficient has been obtained varying the ratio Xr_H/Xr_M as a function of the calibration coefficient (C_{cal}). Taking into account a $C_{cal} = 1.24 \pm 0.04$, the fraction Xr_H/Xr_M approaches to 1. This means that the corrected hindcast time series (by means of C_{cal}) shows equal values to those evaluated on the basis of observed time series.

In order to gain insight on the ECDF of the observed and hindcast extreme values, the Q-Q plots for the uncorrected series (i.e. C_{cal} = 1) and with a correction equal to 1.28 (see **Figure** 7) have been evaluated. **Figure** 7 shows the results and exhibits the usefulness of using the calibration coefficient in improving the reliability of the hindcast.

In addition, the root-mean-square error (*RMSE*) the *Bias*, the correlation coefficient (R), the index of agreement (d) and the Nash-Sutcliffe efficiency coefficient (*NSE*) have been calculated on the sample of the quantiles of the ECDF of the hindcast and observed extreme values. The *RMSE*, the *Bias* and R are commonly

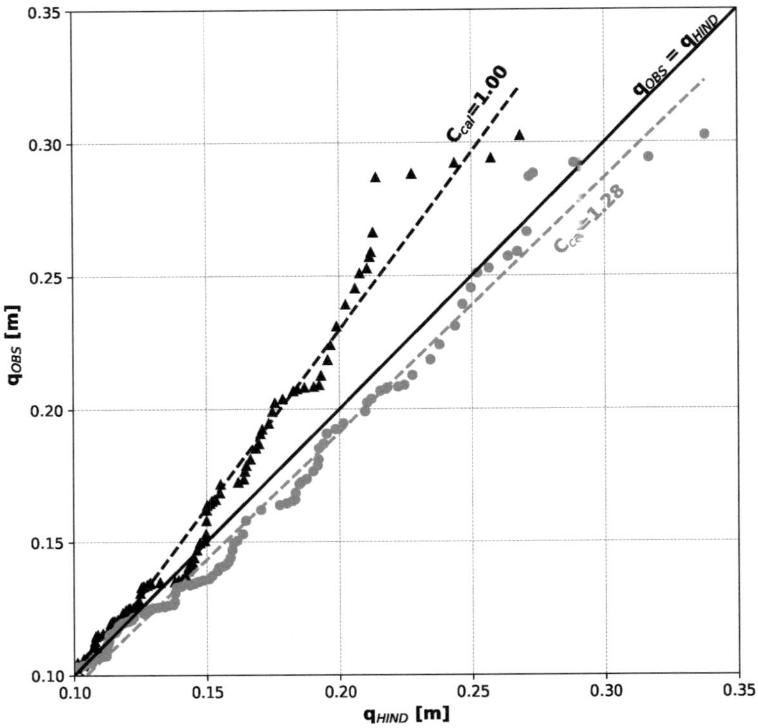

Figure 7.
Q-Q plots for the uncorrected series (black triangle) and with a correction equal to 1.28 (grey circles) of the ECDF quantiles of the extracted extreme values.

used in the literature (e.g. [59–61]), while *d* and *NSE* are less. The index of agreement (e.g. [62]) measures the model error and varies between 0 (no accordance) and 1 (perfect agreement). Instead, the Sutcliffe efficiency coefficient is widely used to assess the goodness of a fit (e.g. [63]), and its values range from $-\infty$ up to 1 (perfect agreement).

These statistical indicators have been calculated considering the corrected (C_{cal} = 1.28) and uncorrected (C_{cal} = 1) hindcast data.

Results show that there is a moderate increase in the reliability of the corrected data (by means of the calibration coefficient). More in details, the *RMSE* ranged from 0.020 to 0.010, the *Bias* has varied from −0.012 to 0.007, *R* varied from 0.987 to 0.991, *d* changed from 0.952 to 0.990, and *NSE* has varied from 0.853 to 0.960. Remembering the aim of the method, however, the same indexes have been evaluated for the quantiles greater than 0.15 m. In this case, the field contains real extreme values, and the importance of the correction is emphasized (*RMSE*: 0.030 → 0.012, *Bias*: −0.025 → 0.009, *R*: 0.975 → 0.982, *d*: 0.849 → 0.977, *NSE*: 0.518 → 0.912).

4. Conclusions

A simplified real-time forecast method and a simplified method for the estimation of return levels of storm surge in semi-enclosed basins are proposed. Both the two approaches are mixed. Indeed, results coming from a physics-based approach are corrected by means of statistical corrections.

In both cases, the strategy is to estimate the dynamic response function of the basin to a unit wind stress. These functions may be used, following the theory of linear dynamic systems, to compute the response of a considered semi-enclosed basin using whatever wind time series.

In this way, only the wind field role is considered, and the pressure field is not. In order to take into account all the meteorological parameters inducing the storm surge, a statistical correction for both models is proposed.

In the case of forecast models, the statistical correction have been made using a series of artificial neural networks trained with (a) the residual raw level time series, (b) recent residual level estimated at the POI on the basis of the measurements collected during 24 h before the forecast time and (c) the forecasted pressures along the basin as input neurons.

For the hindcast method, instead, the pressure field has been considered using the pressure anomaly and operating a statistical correction using a calibration coefficient.

The approaches allow to reduce the computational costs since the numerical simulations have to be done once and for all for each considered basin.

The two methods have been applied to two different points of interest in the Adriatic Sea revealing in both cases good reliability of the obtained results compared to their simplicity.

It has to be noticed that these approaches are devoted to study storm surges in the semi-enclosed basins and are not able to correctly reproduce storm surges due to very rapid meteorological events (i.e. hurricanes).

Acknowledgements

The author wants to thank Prof. Marcello Di Risio for his help in suggesting ideas and methods that have made possible the realization of these approaches. He is also grateful to all the co-authors in the papers regarding storm surges that have been cited in the references of this chapter.

Conflict of interest

The author declares no conflict of interest.

Author details

Davide Pasquali
Environmental and Maritime Hydraulic Laboratory (LIam), University of L'Aquila, Civil, Construction-Architectural and Environmental Engineering (DICEAA), L'Aquila, Italy

*Address all correspondence to: davide.pasquali@univaq.it

References

[1] Syvitski JP, Vörösmarty CJ, Kettner AJ, Green P. Impact of humans on the flux of terrestrial sediment to the global coastal ocean. Science. 2005; **308**(5720):376-380

[2] Oppenheimer M, Glavovic B, Hinkel J, van de Wal R, Magnan AK, Abd-Elgawad A, Cai R, et al. Sea Level Rise and Implications for Low Lying Islands, Coasts and Communities. Ocean Sciences Meeting; 2019

[3] Di Risio M, Bruschi A, Lisi I, Pesarino V, Pasquali D. Comparative analysis of coastal flooding vulnerability and hazard assessment at national scale. Journal of Marine Science and Engineering. 2017;**5**(4):51

[4] Salman A, Lombardo S, Doody P. Living with Coastal Erosion in Europe: Sediment and Space for Sustainability. EUCC; 2004

[5] Saponieri A, Valentini N, Di Risio M, Pasquali D, Damiani L. Laboratory investigation on the evolution of a sandy beach nourishment protected by a mixed soft–hard system. Water. 2018; **10**(9):1171

[6] Newton I, Halley E. Philosophiae naturalis principia mathematica, vol. 62. Jussu Societatis Regiae ac typis Josephi Streater, prostant venales apud Sam; 1780

[7] Darwin GH. A general article in the tides. In: Admiralty Scientific Manual. London: The Admiralty; 1886. pp. 53-91

[8] Doodson AT. The harmonic development of the tide-generating potential. Proceedings of the Royal Society of London, Series A. 1921; **100**(704):305-329

[9] McCully JG. Beyond the Moon: A Conversational Common Sense Guide to Understanding the Tides. World Scientific; 2006

[10] Godin G. The analysis of tides and currents, In: Parker BB, editor. Tidal Hydrodynamics. 1991. pp. 675-709

[11] Pugh DT. Tides, Surges and Mean Sea-Level. John Wiley & Sons Ltd.; 1996 (Reprinted with Corrections)

[12] Gonnert G, Dube S, Murty T, Siefert W. Global storm surges: Theory, observations and applications. In: Die Küste, 63 Global Storm Surges. 2001. pp. 623

[13] Horsburgh K, de Vries J, Etala M, Murty T, Seo J, Dube S, et al. Publications, presentations and other activities. In: Guide to Storm Surge Forecasting. 2011th ed. Wmo Publication; 2011

[14] Canestrelli P, Mandich M, Pirazzoli P, Tomasin A. Venti, depressioni e sesse: perturbazioni delle maree a venezia (1951–2000). Centro Previsioni e Segnalazioni Maree; 2001 in Italian

[15] Cerovečki I, Orlić M, Hendershott MC. Adriatic seiche decay and energy loss to the Mediterranean. Deep Sea Research Part I: Oceanographic Research Papers. 1997;**44**(12):2007-2029

[16] Vilibić I. The role of the fundamental seiche in the Adriatic coastal floods. Continental Shelf Research. 2006;**26**(2):206-216

[17] Codiga DL. Unified tidal analysis and prediction using the UTide Matlab functions. In: Technical Report 2011–01. Narragansett, RI: Graduate School of Oceanography, University of Rhode Island; 2011

[18] Pawlowicz R, Beardsley B, Lentz S. Classical tidal harmonic analysis including error estimates in matlab using t_tide. Computers & Geosciences. 2002;**28**(8):929-937

[19] Bajo M, Zampato L, Umgiesser G, Cucco A, Canestrelli P. A finite element operational model for storm surge prediction in Venice. Estuarine, Coastal and Shelf Science. 2007;**75**(1): 236-249

[20] Bernier NB, Thompson KR. Deterministic and ensemble storm surge prediction for Atlantic Canada with lead times of hours to ten days. Ocean Modelling. 2015;**86**:114-127

[21] Daniel P, Haie B, Aubail X. Operational forecasting of tropical cyclones storm surges at meteo-France. Marine Geodesy. 2009;**32**(2): 233-242

[22] Flather RA. Existing operational oceanography. Coastal Engineering. 2000;**41**(1):13-40

[23] Heaps N. Storm surges, 1967–1982. Geophysical Journal International. 1983; **74**(1):331-376

[24] Harris DL. The equivalence between certain statistical prediction methods and linearized dynamical methods. Monthly Weather Review. 1962;**90**(8): 331-340

[25] Vilibić I, Šepić J, Pasarić M, Orlić M. The Adriatic Sea: A long-standing laboratory for sea level studies. Pure and Applied Geophysics. 2017;**174**(10): 3765-3811

[26] Bargagli A, Carillo A, Pisacane G, Ruti P, Struglia M, Tartaglione N. An integrated forecast system over the Mediterranean basin: Extreme surge prediction in the northern Adriatic Sea. Monthly Weather Review. 2002;**130**(5): 1317-1332

[27] Lionello P, Zampato L, Malguzzi P, Tomasin A, Bergamasco A. On the correct surface stress for the prediction of the wind wave field and the storm surge in the northern Adriatic Sea. Il Nuovo Cimento C. 1998;**21**(5):515-532

[28] Bertotti L, Bidlot J-R, Buizza R, Cavaleri L, Janousek M. Deterministic and ensemble-based prediction of Adriatic Sea sirocco storms leading to 'acqua alta'in Venice. Quarterly Journal of the Royal Meteorological Society. 2011;**137**(659):1446-1466

[29] Canestrelli E, Canestrelli P, Corazza M, Filippone M, Giove S, Masulli F. Local learning of tide level time series using a fuzzy approach. In: 2007 International Joint Conference on Neural Networks, IEEE. 2007. pp. 1813-1818

[30] Michelato A, Mosetti R, Viezzoli D. Statistical forecasting of strong surges and application to the lagoon of Venice. Bollettino di Oceanologia Teorica ed Applicata.1983;**1**:67-83

[31] Petaccia S, Serravall R, Pellicano F. Improved method of sea level forecasting at Venice (northern Adriatic Sea). Communications in Nonlinear Science and Numerical Simulation. 2006;**11**(3):281-296

[32] Tomasin A, Frassetto R, Nihoul J. Cyclogenesis and forecast of dramatic water elevations in Venice. Marine Forecasting. 1979:427-437

[33] Tosoni A, Canestrelli P. Il modello stocastico per la previsione di marea a venezia. Atti Ist Veneto Sci Lett Arti. 2011;**169**:2010-2011

[34] Bajo M, Umgiesser G. Storm surge forecast through a combination of dynamic and neural network models. Ocean Modelling. 2010;**33**(1):1-9

[35] Mel R, Lionello P. Probabilistic dressing of a storm surge prediction in the Adriatic Sea. Advances in Meteorology. 2016;**2016**. DOI: 10.1155/2016/3764519

[36] Pasquali D, Di Risio M, De Girolamo P. A simplified real time method to forecast semi-enclosed basins

storm surge. Estuarine, Coastal and Shelf Science. 2015;**165**:61-69

[37] Molteni F, Buizza R, Palmer TN, Petroliagis T. The ecmwf ensemble prediction system: Methodology and validation. Quarterly Journal of the Royal Meteorological Society. 1996; **122**(529):73-119

[38] Yukimoto S, Adachi Y, Hosaka M, Sakami T, Yoshimura H, Hirabara M, et al. A new global climate model of the meteorological research institute: Mri-cgcm3—Model description and basic performance. Journal of the Meteorological Society of Japan. Ser. II. 2012;**90**:23-64

[39] Pasquali D, Bruno M, Celli D, Damiani L, Di Risio M. A simplified hindcast method for the estimation of extreme storm surge events in semi-enclosed basins. Applied Ocean Research. 2019;**85**:45-52

[40] Di Risio M, Pasquali D, Lisi I, Romano A, Gabellini M, De Girolamo P. An analytical model for preliminary assessment of dredging-induced sediment plume of far-field evolution for spatial non homogeneous and time varying resuspension sources. Coastal Engineering. 2017;**127**:106-118

[41] Roos PC, Schuttelaars HM. Influence of topography on tide propagation and amplification in semi-enclosed basins. Ocean Dynamics. 2011; **61**(1):21-38

[42] Haidvogel DB, Arango H, Budgell WP, Cornuelle BD, Curchitser E, Di Lorenzo E, et al. Ocean forecasting in terrain-following coordinates: Formulation and skill assessment of the regional ocean modeling system. Journal of Computational Physics. 2008;**227**(7): 3595-3624

[43] Hervouet J, Ata R. User manual of opensource software telemac-2d. In: Report. Paris: EDF-R&D; 2017

[44] De Girolamo P, Di Risio M, Beltrami G, Bellotti G, Pasquali D. The use of wave forecasts for maritime activities safety assessment. Applied Ocean Research. 2017;**62**:18-26

[45] Cecconi G, Canestrelli P, Corte C, Di Donato M. Climate record of storm surges in Venice. In: RIBAMOD–River Basin Modelling, Management and Flood Mitigation, Concerted Action, Proceedings of the final Wallingford Workshop, vol. 18287. Luxembourg, EUR: European Commission; 1999. pp. 149-156

[46] Schwab DJ, Rao DB. Barotropic oscillations of the Mediterranean and Adriatic seas. Tellus A. 1983;**35**(5): 417-427

[47] You SH, Lee W-J, Moon KS. Comparison of storm surge/tide predictions between a 2-d operational forecast system, the regional tide/storm surge model (rtsm), and the 3-d regional ocean modeling system (roms). Ocean Dynamics. 2010;**60**(2):443-459

[48] Amante C, Eakins B. Arc-Minute Global Relief Model: Procedures, Data Sources and Analysis (Etopo1). Boulder, Colorado, USA: NOAA, National Geophysical Data Center; 2009

[49] Drago M, Iovenitti L. σ-Coordinates hydrodynamic numerical model for coastal and ocean three-dimensional circulation. Ocean Engineering. 2000; **27**(10):1065-1085

[50] Garratt J. Review of drag coefficients over oceans and continents. Monthly Weather Review. 1977;**105**(7): 915-929

[51] Defant A. Physical Oceanography. Vol. 1. Pergamon; 1961

[52] Celli D, Pasquali D, De Girolamo P, Di Risio M. Effects of submerged berms on the stability of conventional rubble mound breakwaters. Coastal Engineering. 2018;**136**:16-25

[53] Di Risio M, Lisi I, Beltrami G, De Girolamo P. Physical modeling of the cross-shore short-term evolution of protected and unprotected beach nourishments. Ocean Engineering. 2010;**37**(8–9):777-789

[54] Dee DP, Uppala S, Simmons A, Berrisford P, Poli P, Kobayashi S, et al. The era-interim reanalysis: Configuration and performance of the data assimilation system. Quarterly Journal of the Royal Meteorological Society. 2011;**137**(656):553-597

[55] Apulia Region. Apulia region meteomarine network. 2018. Available from: http://www.adb.puglia.it/public/news.php?extend.282.6 [Accessed: 06 May 2018]

[56] Damiani L, Bruno MF, Molfetta MG, Nobile B. Coastal zone monitoring in Apulia region: First analysis on meteomarine climate. In: Proceedings of 5th International Symposium on Environmental Hydraulics (ISEH 2007); Arizona, USA. 2007

[57] Samayam S, Laface V, Annamalaisamy SS, Arena F, Vallam S, Gavrilovich PV. Assessment of reliability of extreme wave height prediction models. Natural Hazards and Earth System Sciences. 2017;**17**(3):409-421

[58] Coles S, Bawa J, Trenner L, Dorazio P. An Introduction to Statistical Modeling of Extreme Values. Vol. 208. Springer; 2001

[59] Durrant TH, Greenslade DJ. Spatial evaluations of access marine surface winds using scatterometer data. Australian Meteorlogical and Oceanographic Journal. 2012;**62**:263-276

[60] Moriasi DN, Arnold JG, Van Liew MW, Bingner RL, Harmel RD, Veith TL. Model evaluation guidelines for systematic quantification of accuracy in watershed simulations. Transactions of the ASABE. 2007;**50**(3):885-900

[61] Willmott CJ, Robeson SM, Matsuura K. A refined index of model performance. International Journal of Climatology. 2012;**32**(13):2088-2094

[62] Willmott CJ, Ackleson SG, Davis RE, Feddema JJ, Klink KM, Legates DR, et al. Statistics for the evaluation and comparison of models. Journal of Geophysical Research: Oceans. 1985;**90**(C5):8995-9005

[63] McCuen RH, Knight Z, Cutter AG. Evaluation of the Nash–Sutcliffe efficiency index. Journal of Hydrologic Engineering. 2006;**11**(6):597-602

Chapter 8

Physical and Numerical Modeling of Landslide-Generated Tsunamis: A Review

Alessandro Romano

Abstract

Landslide-generated tsunamis represent a serious source of hazard for many coastal and lacustrine communities. The understanding of the complex physical phenomena that govern the tsunami generation, propagation and interaction with the coast is essential to reduce and mitigate the tsunamis risk. Experimental, analytical, and numerical models have been extensively used (both as separated tools and in conjunction) to shed light on these complicated natural events. In this work, a non-exhaustive update of the state of the art related to the physical and numerical modeling techniques of landslide-generated tsunamis, with a special focus on those studies published in the last ten years, is provided. As far as numerical models are concerned, a special attention is paid to the most recently developed Computational Fluid Dynamics (CFD) techniques, whose development and application have experienced a boost up the last decade.

Keywords: landslide-generated tsunamis, physical modeling, numerical modeling, computational fluid dynamics (CFD), water waves

1. Introduction

Impulsive waves (i.e. tsunamis) are likely to be generated by earthquakes, landslides, volcanic eruptions, impacts of asteroids and gradients of atmospheric pressure (Løvholt et al. [1]). There are coastal areas which are particularly prone to landslide-generated tsunami risk. The destructive effects caused by the impulsive waves, generated by landslide sources, can be strongly magnified by the characteristics of the so-called "confined geometries" (e.g. bays, reservoirs, lakes, volcanic islands, fjords, etc.). Complicated physical phenomena (e.g. trapping mechanisms, edge waves, wave runup, etc.) take place as a consequence of the interaction between the generated waves and the local bathymetry controlling the tsunami propagation and interaction with the coast. Many past events of landslide-generated tsunamis testify this reality (e.g. Lake Geneva, Switzerland, *Kremer et al.* [2]; Lituya Bay, Alaska, *Fritz et al.* [3]; Vajont Valley, Italy, *Panizzo et al.* [4]; Stromboli Island, Italy, *Tinti et al.* [5]; Papua New Guinea, *Synolakis et al.* [6]; Anak Krakatau, Indonesia, *Grilli et al.* [7]).

Figure 1 provides good examples of areas prone to landslide tsunami hazard (upper left panel: Lituya Bay, Alaska; upper right panel: Vajont Valley, Italy; lower panels: Stromboli Island, Italy). The physical process at hand is generally characterized by smaller length and time scales than those of tsunamis generated by

(a) (b) (c) (d)

Figure 1.
Pictures of areas historically affected by landslide-generated tsunamis. (a) Lituya Bay, Alaska, 1958. (b) Vajont Valley, Italy, 1963. (c) Stromboli Island, Italy, 2002. (d) Damages at Stromboli Island, Italy, 2002.

earthquakes. The triggering mechanism (the landslide), can be classified as subaerial, partially submerged or completely submerged, depending on the initial landslide position [8, 9]. The occurrence of the landslide at the water body boundary implies that the generated impulse waves propagate both seaward and alongshore. Moreover, complicated physical phenomena due to the interaction between the waves and the sea bottom (e.g. trapping mechanisms *Bellotti and Romano* [10], *Romano et al.* [11]) are likely to plays a significant role, which comprehension is essential for designing and implementing the so-called early warning systems *Bellotti et al.* [12], Cecioni et al. [13], De *Girolamo et al.* [14].

The complex physical phenomena related, on one hand, to the landslide triggering mechanisms and, on the other hand, to the tsunami generation, propagation and interaction with the coast mechanisms are brilliantly and exhaustively described by **Figure 1** (and its description) of *Di Risio et al.* [8]. Due to its clarity and usefulness, this figure is here reported (**Figure 2** of the present manuscript). As mentioned, depending on the initial landslide position, as well as on the geometry of the near- and the far-field, the tsunami characteristics can change significantly. Therefore, to reduce and/or to mitigate the landslide-generated tsunami risk the comprehension and the right modeling of such complicated phenomena is essential. Numerous studies dealing with landslide-generated tsunamis are available in the scientific literature. Experimental, analytical, and numerical models have been extensively used (both as separated tools and in conjunction) to shed light on this complicated natural event.

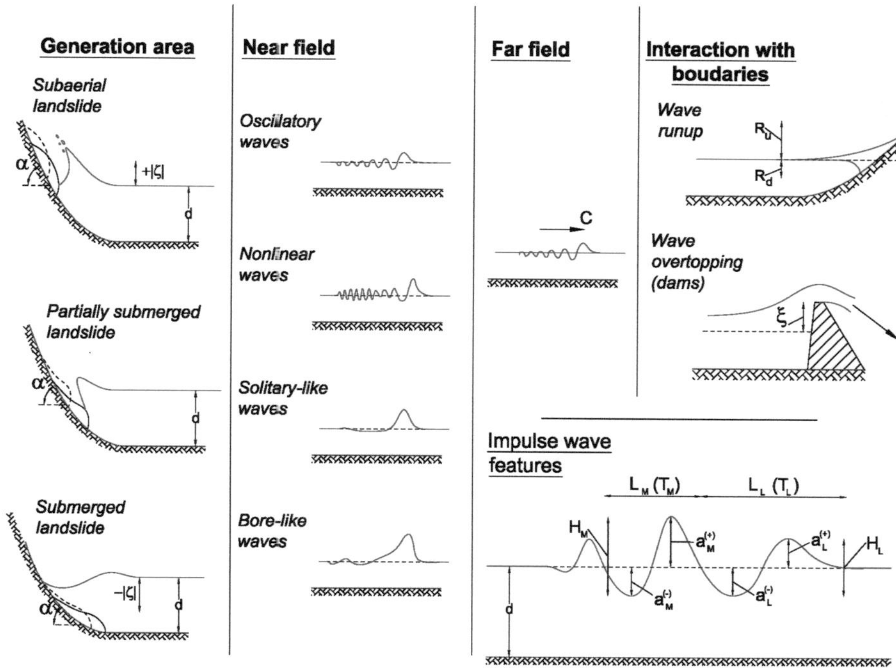

Figure 2.
Sketch of landslide-generated impulse waves [8].

In this chapter a special attention to the experimental and numerical modeling of landslide-generated tsunamis is given. It is important to highlight that this work has not the haughtiness to provide an exhaustive description, since the beginning to the present days, of the physical and numerical modeling techniques related to landslide-generated tsunamis. Indeed, brilliant and exhaustive review of the state of the art, as well as of the future challenges, related to the physical and numerical modeling of landslide-generated tsunamis are provided by authoritative Authors. It is worth to remember, among others, the excellent review of the physical model experiments, together with the main results and achievements, provided *Di Risio et al.* [8], as well as the exhaustive description of the landslide-generated tsunami numerical modeling techniques addressed by *Yavari-Ramshe and Ataie-Ashtiani* [15]. Thus, the objective of the present chapter lies in providing an update of the state of the art related to the physical and numerical modeling techniques of landslide-generated tsunamis, with a special focus on those studies published in the last ten years. Moreover, as far as numerical models are concerned a special attention is paid to the recently developed Computational Fluid Dynamics (CFD) techniques; in fact, the development and the application of these techniques has experienced a boost up the last decade. It is worth noticing that analytical modeling is not considered in the Chapter.

As stated, in this study a review of the physical and numerical modeling techniques related to landslide-generated tsunamis is provided. The main purpose of this study lies in describing, with no claim to be exhaustive and by adopting a flowing style, the main approaches exploited so far and in discussing the potentials as well as the limitation of the methods themselves. Moreover, the future challenges related to the present research field are discussed. This chapter in organized as follows. In the next section a review of the physical modeling techniques related to landslide-generated tsunamis is presented. Then, a section dealing with the numerical modeling techniques follows. Finally, concluding remarks close the chapter.

2. Landslide-generated tsunamis: physical models

In this section the most recent physical model experiments related to landslide-generated tsunamis, dealing with a large variety of geometries (plane slopes, conical islands, reservoirs, etc.) and landslide types (subaerial, partially and completely submerged), are reported. As anticipated, the studies are enumerated using a flowing style and presented in chronological order of publication. A distinction between rigid and deformable landslide models has been followed, while no distinction is made between 2D and 3D configurations.

2.1 Rigid landslide models

In 2013, *Romano et al.* [11] carried out a 3D experiment to investigate the alongshore propagation features and the trapping mechanisms of tsunamis generated by semi-elliptical subaerial landslides around the coast of a circular island. They used the same experimental setup described by *Di Risio et al.* [16] and later by *Romano et al.* [17] and, by applying the wavenumber-frequency analysis (k-f) on the records of shoreline displacement, they pointed out that the 0th-order edge wave mode is the only one relevant for shoreline runup.

Heller and Spinneken [18] performed a large number of 2D experiments in a wave flume dealing with subaerial block-shaped landslides. In their experiments they investigated, among others, the effect of three block model parameters (i.e. the landslide Froude number, the relative slide thickness, and the relative slide mass). They provided empirical equations for the maximum wave amplitude, height, and period. Moreover, by comparing the newly derived equations as obtained for block-shaped landslides with the available equations for granular landslides, they found that block-shaped landslides do not necessarily generate larger waves than granular slides.

In 2015, *Heller and Spinneken* [19] presented a new set of 3D experiments carried out in a wave tank dealing with subaerial block-shaped landslides. The authors compared the new 3D results with past 2D experimental ones, published by the same pair of Authors [18], taking advantage of the identical boundary conditions between the two sets of data. Several parameters (i.e. water depth, landslide volume and density, landslide release positions) have been changed during the new experiments. Therefore, the Authors provided some empirical equations to predict the 3D offshore and laterally onshore wave properties, identifying the waves decay law both for 2D and 3D configurations, and providing very useful discussion on the existing 2D-3D conversion formulae (e.g. *Watts et al.* [20])

In 2016, *Romano et al.* [17] published a series of new 3D experiments dealing with tsunamis generated by semi-elliptical subaerial landslides occurring at the flank of conical islands. As pointed out by *Di Risio et al.* [16], the physical model at hand aims at reproducing, in a Froude law scale, the Sciara del Fuoco slope (see **Figure 3**), i.e. a natural sliding surface located at the Stromboli Island (Southern Thyrrenian Sea, Italy). The main objective of the experiments is to provide a benchmark dataset for the validation of numerical models of landslide-generated tsunamis. To this end, a quite unique acquisition system, consisting of both fixed and movable wave gauges, has been deployed and used. The experimental procedure is per se a novelty as each experiment consists in repeating several times the same landslide event, by changing for each repetition the position of the movable gauges, then obtaining, after checking the repeatability, a single virtual experiment with high spatial resolution measurements. Two different semi-elliptical landslide bodies have been used for the experiments (see **Figure 3**) in order to investigate the effects of the landslide volume and thickness, revealing that the mentioned parameters affects significantly only the wave amplitudes, especially in the near field,

Figure 3.
Pictures of the Sciara del Fuoco (Stromboli island) and the slide placed on the conical island model. (a) Detail of Sciara del Fuoco: rock fall. (b) Detail of the slide along the conical island. (c) Aerial view of Sciara del Fuoco. (d) Lateral view of the conical island and the slide.

while for the wave periods (and celerities), a weak dependence upon these landslide parameters can be observed.

The studies cited so far are related to subaerial landslides. Few are the recent experimental works dealing with submerged landslides. Indeed, the experimental modeling of submerged landslide presents a wide range of physical restrictions. Therefore, clever technical solutions have often been employed (e.g. *Enet and Grilli* [21], *Liu et al.* [22], Watts [23]). Nevertheless, it remains the practical difficulty of exploring in detail the influence of some key governing parameters, as for instance the initial acceleration a_0. This parameter is commonly recognized to be a crucial one in the slide kinematics, in particular in the initial phase, when the energy transfer between the landslide and the water takes place [1, 20, 21, 23–26]. Several experimental studies explored the importance of a_0 by means of different techniques. *Watts* [23] changed the landslide's density to obtain different values of a_0. In 2017, *Romano et al.* [27], using the same physical model described by *Di Risio et al.* [16], *Romano et al.* [11, 17], used a mechanical system controlled by an electric motor to perform parametric 3D experiments by changing the kinematics of a semi-elliptical submerged landslide. The experimental results pointed out that as the initial acceleration increases, then the rising time of the first wave trough decreases and, in general, the wave signals exhibit a shorter wave periods. This findinds have been recently numerically confirmed by *Romano et al.* [28] (see Section 3). As far as submerged landslides are concerned, a note on the use of rigid landslide models is due. Although this represents an approximation of the real submerged landslide behavior, it is well demonstrated in the scientific literature (e.g. *Grilli et al.* [24]) that the landslide deformation does not play a significant role on submarine landslide tsunami features in the slide early time kinematics, which at short time scales is mainly governed by the initial acceleration.

Finally, innovative physical model approaches, by using rigid landslide models, have been recently used by *Perez del Postigo Prieto et al.* [29] to reproduce a coupled-source tsunami generation mechanism due to a 2D underwater fault rupture followed by a submarine landslide. Furthermore, 2D experiments have been employed to interpret the dynamics of complicated recent events, like the eruption of the Anak Krakatoa volcano (Indonesia) in December 2018. To this end, it is worth noticing the study of *Heidarzadeh et al.* [30], that applied a combination of qualitative physical modeling and wavelet analyses of the tsunami as well as numerical modeling to propose a source model of the Anak Krakatoa event.

2.2 Deformable landslide models

Quite various are the physical model tests dealing with deformable landslide models. In 2010 Heller and Hager [31] performed 2D experiments dealing with subaerial landslide by using granular slide material. The large number of tests (more than 200) aimed at exploring the influence of several governing parameters, namely: still water depth, slide impact velocity, slide thickness, bulk slide volume, bulk slide density, slide impact angle, and grain diameter. As a result, the Authors provided empirical predictive equations for all relevant wave characteristics, e.g. maximum wave height, the maximum wave amplitude (including its location and period in the slide impact zone), and both the wave height and amplitude decay and the period increase in the wave propagation zone. Furthermore, the Authors present a comparison of the presented equations with the 1958 Lituya Bay case, finding a good agreement.

In 2012, *Mohammed and Fritz* [32] carried out a massive 3D experimental campaign aimed at studying the tsunamis characteristics generated by subaerial deformable granular landslides occurring at a plane slope, using a novel approach based on a pneumatic landslide generator to control the landslide impact characteristics. In their study they found a robust correlation between the wave characteristics and the landslide Froude number, providing also some quantitative calculation of the landslide-water energy converted rate. Moreover, a deep discussion on the wave amplitudes decay, wave celerities and comparison with other 2D and 3D landslide tsunami studies is provided.

In 2014, Viroulet et al. [33] published the results of some 2D experiments dealing with subaerial landslides sliding along a rough slope. The Authors investigated mainly the influence of the slope angle and the granular material, by using three different granular materials (spherical glass beads with two different diameter, non-spherical sand), on the initial amplitude of the generated leading wave and the evolution of its amplitude during the propagation. Interestingly, the presented experiments aim at investigating the tsunami characteristics generated by landslide characterized by Froude number smaller than one. As stated by the Authors, this situation is particularly relevant to model tsunamis generated by cliff failures located just above the sea surface, which are characterized by low impact velocities.

A unique series of large-scale 3D physical model experiments is described in 2016 by *McFall and Fritz* [9]. In this study the Authors investigated the runup features, measured both on the same coast at which the landslide occurred and on an opposing hill slope, of tsunamis generated by subaerial granular landslides occurring both at planar coast and conical island. The pneumatic landslide generator described in Mohammed and Fritz [32] has been used for the experiments. Different landslides geometries and kinematics have been used and robust results and findings are provided, namely related to: maxima and minima runup and rundown location, decay along the coast and amplification; effects of the granulometry on the lateral wave runup; energy trapping properties of a circular shoreline, also confirming the findings of *Di Risio et al.* [16], *Romano et al.* [11, 17]. Finally, predictive equations for the laterally propagating wave characteristics, benchmarked against the 2007 landslide-generated tsunami in Chehalis Lake, British Columbia, Canada, are provided by the Authors.

In the same year, *Lindstrøm* [34] performed a series of 2D experiments dealing with subaerial landslides in a wave flume. The Author, keeping constant few parameters (i.e. landslide volume, initial position, slope angle and equilibrium water depth), varied only the slide material. In particular, five different slide types have been used: one block slide and four granular slides with grain diameter ranging from 3 mm to 25 mm. A very interesting aspect, pointed out by the present study, is related to the effect of the landslide porosity. Indeed, *Lindstrøm* [34] by comparing the present results with the predictive formulae of maximum wave amplitudes, available in the literature, found some differences, probably due to the effects of the landslide permeability.

Also in 2016, *Zitti et al.* [35] carried out a series of 2D experiments dealing with subaerial landslides, aiming at simulating the effects of a snow avalanche entering a body of water. To mimic a snow avalanche striking a reservoir, a lightweight granular material has been used as a substitute for snow. Morevoer, the Authors developed a theoretical model to describe the momentum transfers between the particle and water phases of such events. The presented experimental results have also been compared with those obtained by *Heller and Hager* [31], as the same relative particle density, but higher landslide Froude numbers, has been used by the two groups of Authors.

In 2017 *Miller et al.* [36] carried out 2D experiments dealing with granular subaerial landslides. The Authors presented a detailed analysis on the velocity and thickness of the granular flow, on the shape and location of the submarine landslide deposit, on the amplitude and shape of the near-field wave, on the far-field wave evolution, and on the wave runup elevation on a smooth impermeable slope. By using high-speed camera observations and standard free surface elevation measurements the Authors pointed out that only a portion of the landslide (named the "effective mass") is engaged in activating the leading wave. Furthermore, the Authors observed a good agreement between their experimental results and the values provided by existing empirical predictive formulae, available in the literature, as the so-called effective mass is used. The effective mass is defined as the percentage of the total landslide mass that enters the water body before the initial wave leaves the impact zone and it is a crucial aspect to be considered for landslides that are long and thin with very large relative mass, as in this case only a portion of the landslide mass is engaged in activating the leading wave. In the same year, *Mulligan and Take* [37], by using the experimental data discussed by *Miller et al.* [36], presented a study on the momentum flux exchange between granular landslides and water, finding that the results of their approach, based on the momentum-based equations, are in agreement with the previous laboratory data of *Heller and Hager* [31] and *Miller et al.* [36].

In the last two years, peculiar and very interesting new experimental approaches have been used. It is worth citing the study of *Tang et al.* [38] that performed 2D laboratory tests for impulse waves generated by subaerial landslides made as the combination of solid block and granular materials (glass spheres), also comparing the obtained results with those of individual models of pure solid block and granular landslides. In their experiments the Authors varied the slope angle and the mass ratio m^* (i.e. mass of the solid block divided by mass of the granular material). The experimental results suggest that the mixed landslide composition generally produces larger impulse waves in the impact zone compared with those triggered by pure solid block landslides and pure granular landslides, suggesting that the primary wave amplitudes of impulse waves might have been underestimated in previous laboratory tests with solely solid or granular assemblies when using the same slide mass and release height. Furthermore, they pointed out that, if compared with pure granular landslides, the combined landslides generally exhibit larger Froude numbers and slide thickness.

Very recently, *Bullard et al.* [39] performed 2D tests dealing with deformable subaerial landslides. The point of novelty, among others, of the present study lies in using water as sliding material. This aspect ensures a null internal shear strength, being then representative of the upper limit of high landslide mobility. Four different slide volumes have been used during the experiments and a high-speed camera has been used to measure the slide thickness and velocity. The experimental results indicate that in the near-field the maximum wave amplitude is dependent on the landslide thickness and velocity and is relatively independent of the water depth.

3. Landslide-generated tsunamis: numerical models

Although totally irreplaceable, experimental tests are often time consuming, especially if 3D models are considered. Large facilities, as well as complex experimental configurations and sophisticated measurement systems are often needed (see [9, 17]). Furthermore, it is not always possible to explore in detail the influence of all the involved parameters. In this sense, tsunamis generated by submerged landslides provide a good example. Often the waves generated by submerged landslides are too small to get reliable measurements in the experimental facilities. Moreover, as previously stated, it can be difficult to explore the influence of key governing parameters (e.g. the initial acceleration a_0, *Romano et al.* [27]).

In this sense, numerical modeling can provide a valuable complementation to the physical model experimental activities. Indeed, numerical modeling techniques have progressively supported physical ones in shedding light on the complex physical phenomena involved in the generation and propagation mechanisms of landslide-generated tsunamis. Similarly to experimental models, a multitude of approaches has been adopted during recent years for numerically modeling landslide-generated tsunamis (an extensive review has been provided by *Yavari-Ramshe and Ataie-Ashtiani* [15]). Eulerian and Lagrangian frameworks with three grid types (structured, unstructured, and meshless) have been used for tsunami simulations, employing both depth-averaged models, using Non-Linear Shallow Water or Boussinesq Equations, and Navier–Stokes models, considering both 2D and 3D configurations (e.g. [7, 13, 22, 25, 30, 40–47]).

The most recently developed tools offered by Computational Fluid Dynamics (CFD) can provide a significant support for shedding light on many of the unresolved aspects. In particular, they can be very useful to model the near-field wave characteristics. Indeed, the accurate reproduction of the momentum exchange between the landslide and the water body, achievable by the CFD methods, is crucial for a detailed modeling of tsunami generation, propagation and the interaction with the coastline.

In this last section of the chapter, a brief overview of the studies dealing with the recent CFD techniques and approaches developed and published in the last ten years is presented.

In 2010, *Abadie et al.* [48] presented the application and the experimental validation of the 3D incompressible multiple-fluid Navier–Stokes Volume Of Fluid (VOF) model THETIS to reproduce waves generated by rigid and deforming landslides valid for idealized geometries. All the domain portions (i.e. water, air, and landslide) are treated as Newtonian fluids. In this case, as far as rigid slides are concerned, a "penalty method" allows for parts of the fluid domain to behave as a solid. Thus, the coupling between a rigid slide and water is implicitly computed and it is not necessary to specify a given landslide kinematics. The comparison between numerical model simulations and experimental results, related to different landslide configuration (semi-elliptical block, vertical falling rectangular block and 2D and 3D wedges sliding down an incline), shows a good agreement.

One year later, *Montagna et al.* [49] carried out some 3D numerical computations of landslide-generated tsunamis by using the commercial code FLOW-3D, dealing with a semi-elliptical rigid subaerial landslide occurring at the coast of a conical island. A very good agreement is found by comparing the numerical runup measurements with the experimental data obtained by *Di Risio et al.* [16].

In 2015, *Ma et al.* [50] described a new two-layer model for subaerial granular landslide motion and tsunami wave generation. In this study, the modeling of the landslide motion and tsunami wave generation are simulated by separate model components. Indeed, the landslide is described as a saturated granular debris flow, accounting for intergranular stresses governed by Coulomb friction. Tsunami wave generation and propagation is simulated by the 3D Non-Hydrostatic WAVE model NHWAVE [51] that solves the incompressible Navier–Stokes equations. It is worth noticing, that the hybrid numerical approaches (i.e. coupling geotechnics and hydrodynamics models) have been successfully carried out by *Løvholt et al.* [42] and later by *Kim et al.* [25], that simulated the dynamics of the Storegga Slide and tsunami using the depth-averaged landslide model BingClaw, which implements visco-plastic rheology and remolding, and couple it to a standard tsunami propagation model, to reproduce tsunamis generated by submerged landslides.

Heller et al. [52] presented a composite (experimental-numerical) modeling approach for modeling tsunamis generated by rigid subaerial landslides. In this case, an hybrid approach based on the combined use of physical and numerical modeling has been used. Indeed, the experimental results, described in *Heller and Spinneken* [19], have been used to calibrate the 3D smoothed particle hydrodynamics (SPH) code DualSPHysics v3.1 [53], which includes a discrete element method (DEM)-based model to simulate the landslide-ramp interaction.

In 2016, *Shi et al.* [54] presented 2D simulations of the generation of impulse waves produced by subaerial granular landslides. They used a newly-developed soil-water coupling model in a smoothed particle hydrodynamics (SPH) framework. The point of novelty of the work lies in using an elasto-plastic constitutive model for soil, a Navier–Stokes equation based model for water, and a bilateral coupling model at the interface. The Authors tested their model with simulated waves induced by both slow and fast landslides, obtaining a good agreement between numerical and experimental data. The experimental benchmark data used by *Shi et al.* [54] to test their model are the ones described by *Viroulet et al.* [33], aiming at reproducing slow landslides, and by *Fritz et al.* [55], to simulate fast landslides. Another strength of their modeling approach is related to that all parameters used in the model have their physical meaning in soil mechanics and can be obtained from conventional soil mechanics experiments directly.

Whittaker et al. [47] presented the 2D physical and numerical modeling of a submerged rigid semi-elliptical block body moving along a horizontal and impermeable surface (i.e. the sea bottom). During the experiments, the body movement was controlled by mechanical system and laser-induced fluorescence measurement systems has been used for measuring both spatial and temporal variations in the free surface elevation. To numerically reproduce the experiments, the Authors used the OpenFOAM® platform [56]. In particular the Authors used IHFOAM [57, 58], a solver based on interFoam of OpenFOAM® that includes wave boundary conditions and porous media solvers for coastal and offshore engineering applications and can solve both three dimensional Reynolds-Averaged Navier-Stokes equations (RANS) and Volume-Averaged Reynolds-Averaged Navier-Stokes equations (VARANS) for two phase flows [59, 60], coupled to the VOF, to model the rigid object as a moving bottom boundary. The Authors noticed an under-prediction between the measured and the simulated wave amplitudes, although the wave phasing is fairly reproduced.

In 2018, *Si et al.* [45] performed a series of 2D simulations dealing with subaerial landslides using an advanced two-phase model for dry granular material intruding into a water body. The water-air interface both within and outside the granular material is captured by the VOF method. The inter-granular stresses are formulated based on a general collisional-frictional law developed for underwater granular flows and a modified k-ε model is adopted to describe the turbulence effect of the ambient fluid. *Si et al.* [45] used their numerical model to reproduce past experiments related to subaerial landslides [33], finding a good agreement between experimental and numerical simulation results.

In 2019, *Kim et al.* [61] presented the validation of the 3D numerical model TSUNAMI3D based on the Navier–Stokes equations and the VOF, by comparing numerical results with a set of subaerial landslide laboratory experiments (e.g. *Mohammed and Fritz* [32]) and with the ones provided by the commercial code FLOW3D. In this model water and landslide material are considered incompressible and mainly treated as Newtonian fluids. Simplified material rheology and key parameters required for modeling subaerial landslides have been used. Furthermore, the validation results confirmed that the 3D numerical models with simplified landslide rheology can be used to understand and reproduce the complex non-linear wave propagation and runup generated by subaerial landslides. This is an important result as very often the major source of uncertainties is related to the landslide rheology and parameters.

In the same year, *Clous and Abadie* [62] presented a detailed analysis of the energy transfer mechanisms between granular landslides and water. They used incompressible Navier-Stokes VOF model THETIS [48] to perform 2D simulations of tsunamis generated by granular landslides (both subaerial and submerged) reproducing the experiments of *Viroulet et al.* [33]. As previously stated, in the THETIS model air and water are considered Newtonian fluids. The landslide is modeled as a Newtonian fluid whose viscosity is adjusted to fit the experimental results. For the subaerial case, the Authors pointed out that the viscosity value, if properly adjusted, can be seen as a very coarse approximation of the more elaborated non-Newtonian $\mu(I)$ rheological law [63].

The last three numerical approaches described in this chapter have been published in 2020. *Mulligan et al.* [64] presented a new numerical approach to simulate impulse waves generated by highly mobile subaerial landslides by using the technique of the Particle Finite Element Method (PFEM). This approach combines a Lagrangian finite element solution with an efficient remeshing algorithm and is capable of accurately tracking the evolving fluid free-surface and velocity distribution in highly unsteady flows. To validate their numerical model the Authors reproduced the experiments carried out by *Bullard et al.* [39], in which the slide material is water, aiming at representing an avalanche or a debris flow with high mobility. The Authors found that the 2D numerical model shows a good agreement with the experimental observations in terms of landslide velocity and thickness, wave time series, maximum wave amplitude, wave speed, and wave shape.

Chen et al. [65] performed 3D simulations, by using the OpenFOAM® platform [56], to reproduce the impulse waves generated by calving iceberg. To this end, they applied the Immersed Boundary Method (IBM) which allows to handle and model large displacements of bodies. Large-scale experiments [66] have been used as a benchmark for validating the numerical simulations.

Finally, *Romano et al.* [28] presented a new 3D numerical method for modeling tsunamis generated by rigid and impermeable landslides in OpenFOAM® [56] based on the Overset mesh technique. The Overset mesh is based on the use of two (or more) domains. The outer one (i.e. background domain) allows the motion of one, or more, inner domain(s) (i.e. moving domain) that contains a rigid body. The mutual exchange of information between the two domains is achieved by interpolation. The advantage of this approach, if compared with other methods available to simulate the interaction between a moving body and one or more fluids in OpenFOAM®, e.g. the Immersed Boundary Method [65, 67] is that the resolution around the moving body is extremely accurate (i.e. body-fitted approach) and, which is even more important, remains constant throughout the simulation. Furthermore, to fit the current requirement of the Overset implementation (i.e. required distance between the moving body and the domain boundaries) the slope, on which the landslide body moves, has been modeled as a porous media with a very low permeability by using the VARANS approach proposed by *del Jesus et al.* [68], *Lara et al.* [69] and Losada et al. [70]. The approach has been successfully validated through the experiments carried out by *Liu et al.* [22]. The new method

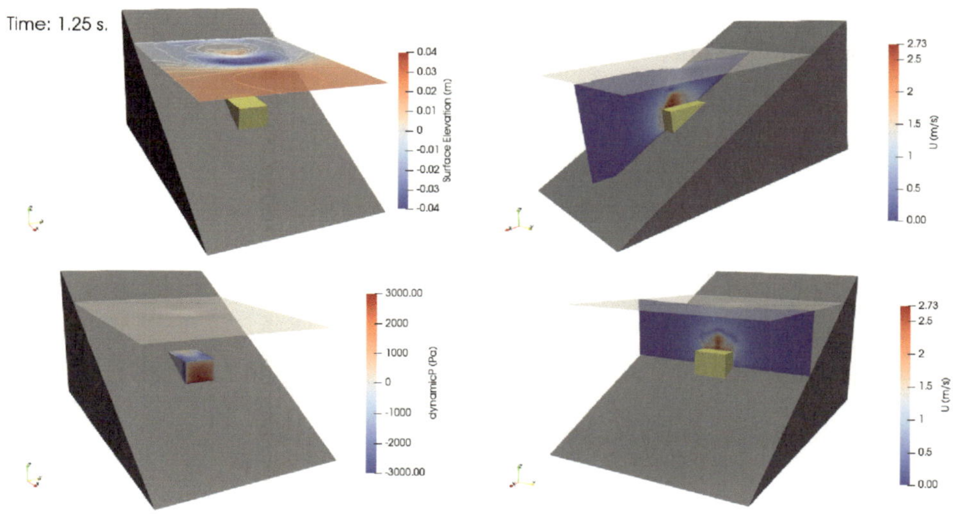

Figure 4.
Contour plot of the free surface elevation (upper left panel), dynamic pressures on the landslide (lower left panel), velocity magnitude on two cross sections (upper and lower right panels) at a given time instant of the numerical simulations described in Romano et al. [28].

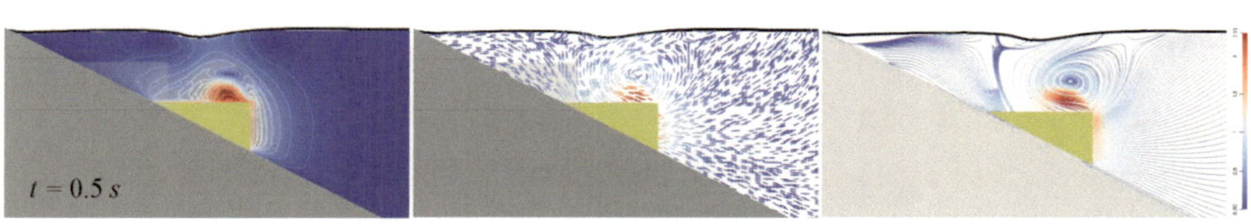

Figure 5.
Velocity magnitude, vectors and streamlines on a cross section at a given time instant of the numerical simulations described in Romano et al. [28].

has then been applied to perform a detailed numerical study of the near-field wave features induced by submerged landslides (see **Figures 4** and **5**), by varying the landslide's initial acceleration a_0. The numerical results, together with previous experimental data [21, 23, 27], have been used to obtain a relationship for predicting the wave properties in the near-field as a function of the Hammack number.

4. Concluding remarks

In this chapter a non-exhaustive update of the state of the art related to the physical and numerical modeling techniques of landslide-generated tsunamis, is presented. As stated, the objective of the present update lies in providing, with no claim to be exhaustive and by adopting a flowing style, the main experimental and numerical, with a special attention to the recently developed Computational Fluid Dynamics (CFD) techniques, approaches published in the last ten years. It is worth to remember that more detailed and complete details on the topic can be found in the detailed reviews of the physical model experiments of the numerical modeling techniques are provided by *Di Risio et al.* [8] and by *Yavari-Ramshe and Ataie-Ashtiani* [15], respectively. As far as the future research challenges are concerned, a special mention to the CFD techniques is due. Indeed, it is well recognized that the models based on the solutions of the Navier-Stokes equations are to be considered as the only alternative to accurately model the tsunami generation process [15]. This kind of modeling approach seems to be mandatory considering the nature of the complex phenomena that govern the impulse waves generation, as confirmed by the achievements obtained by the studies described in the Section 3.

Acknowledgements

A sincere acknowledgment is due to Dr. Marcello Di Risio, Dr. Giorgio Bellotti, Dr. Paolo De Girolamo, Dr. Gabriel Barajas Ojeda and Dr. Javier López Lara for the fruitful discussions on the topic and the constant support.

Conflict of interest

The author declares no conflict of interest.

Author details

Alessandro Romano
Sapienza University of Rome, Rome, Italy

*Address all correspondence to: alessandro.romano@uniroma1.it

References

[1] Løvholt, F., G. Pedersen, C. B. Harbitz, S. Glimsdal, and J. Kim, On the characteristics of landslide tsunamis, *Philosophical Transactions of the Royal Society A: Mathematical, Physical and Engineering Sciences*, *373*, 20140,376, 2015

[2] Kremer, K., G. Simpson, and S. Girardclos, Giant lake geneva tsunami in ad 563, *Nature Geoscience*, *5*, 756–757, 2012

[3] Fritz, H. M., F. Mohammed, and J. Yoo, Lituya Bay landslide impact generated mega-tsunami 50th anniversary, *Pure and Applied Geophysics*, *166*, 153–175, 2009

[4] Panizzo, A., P. De Girolamo, M. Di Risio, A. Maistri, and A. Petaccia, Great landslide events in italian artificial reservoirs, *Natural Hazards and Earth System Sciences*, *5*, 733–740, 2005

[5] Tinti, S., A. Manucci, G. Pagnoni, A. Armigliato, and F. Zaniboni, The 30 December 2002 landslide-induced tsunamis in Stromboli: sequence of the events reconstructed from the eyewitness accounts, *Natural Hazards and Earth System Sciences*, *5*, 763–775, 2005

[6] Synolakis, C. E., J.-P. Bardet, J. C. Borrero, H. L. Davies, E. A. Okal, E. A. Silver, S. Sweet, and D. R. Tappin, The slump origin of the 1998 Papua New Guinea Tsunami, *Proceedings of the Royal Society of London. Series A: Mathematical, Physical and Engineering Sciences*, *458*, 763–789, 2002

[7] Grilli, S. T., et al., Modelling of the tsunami from the December 22, 2018 lateral collapse of Anak Krakatau volcano in the Sunda Straits, Indonesia, *Scientific Reports*, *9*, 2019

[8] Di Risio, M., P. De Girolamo, and G. Beltrami, *Forecasting landslide generated tsunamis: a review*, The Tsunami Threat - Research and Technology, Nils-Axel Marner (Ed.), 2011

[9] McFall, B. C., and H. M. Fritz, Physical modelling of tsunamis generated by three-dimensional deformable granular landslides on planar and conical island slopes, *Proceedings of the Royal Society of London. Series A: Mathematical, Physical and Engineering Sciences*, *472–2188*, 20160,052, 2016

[10] Bellotti, G., and A. Romano, Wavenumber-frequency analysis of landslide-generated tsunamis at a conical island. Part II: EOF and modal analysis, *Coastal Engineering*, *128*, 84–91, 2017

[11] Romano, A., G. Bellotti, and M. Di Risio, Wavenumber–frequency analysis of the landslide-generated tsunamis at a conical island, *Coastal Engineering*, *81*, 32–43, 2013

[12] Bellotti, G., M. Di Risio, and P. De Girolamo, Feasibility of tsunami early warning systems for small volcanic islands, *Natural Hazards and Earth System Sciences*, *9*, 1911–1919, 2009

[13] Cecioni, C., A. Romano, G. Bellotti, M. Di Risio, and P. De Girolamo, Real-time inversion of tsunamis generated by landslides., *Natural Hazards & Earth System Sciences*, *11*, 2011

[14] De Girolamo, P., M. Di Risio, A. Romano, and M. Molfetta, Landslide tsunami: physical modeling for the implementation of tsunami early warning systems in the mediterranean sea, *Procedia Engineering*, 70, 429–438, 2014

[15] Yavari-Ramshe, S., and B. Ataie-Ashtiani, Numerical modeling of subaerial and submarine landslide-generated tsunami waves—recent advances and future challenges, *Landslides*, *13*, 1325–1368, 2016

[16] Di Risio, M., P. De Girolamo, G. Bellotti, A. Panizzo, F. Aristodemo, M. G. Molfetta, and A. F. Petrillo, Landslide-generated tsunamis runup at the coast of a conical island: New physical model experiments, *Journal of Geophysical Research-Oceans*, *114*, 2009b

[17] Romano, A., M. Di Risio, G. Bellotti, M. Molfetta, L. Damiani, and P. De Girolamo, Tsunamis generated by landslides at the coast of conical islands: experimental benchmark dataset for mathematical model validation, *Landslides*, *13*, 1379–1393, 2016

[18] Heller, V., and J. Spinneken, Improved landslide-tsunami prediction: effects of block model parameters and slide model, *Journal of Geophysical Research: Oceans*, *118*, 1489–1507, 2013

[19] Heller, V., and J. Spinneken, On the effect of the water body geometry on landslide–tsunamis: Physical insight from laboratory tests and 2D to 3D wave parameter transformation, *Coast. Eng.*, *104*, 113–134, 2015

[20] Watts, P., S. Grilli, D. Tappin, and G. Fryer, Tsunami generation by submarine mass failure. II: Predictive equations and case studies, *Journal of Waterway, Port, Coastal, and Ocean Engineering*, *131*, 298–310, 2005

[21] Enet, F., and S. T. Grilli, Experimental study of tsunami generation by three-dimensional rigid underwater landslides, *Journal Of Waterway Port Coastal And Ocean Engineering-ASCE*, *133*, 442–454, 2007

[22] Liu, P.-F., T.-R. Wu, F. Raichlen, C. Synolakis, and J. Borrero, Runup and rundown generated by three-dimensional sliding masses, *Journal of Fluid Mechanics*, *536*, 107–144, 2005

[23] Watts, P., Wavemaker curves for tsunamis generated by underwater landslides, *Journal of Waterway, Port, Coastal, and Ocean Engineering*, *124*, 127–137, 1998

[24] Grilli, S. T., O.-D. S. Taylor, C. D. Baxter, and S. Maretzki, A probabilistic approach for determining submarine landslide tsunami hazard along the upper east coast of the united states, *Marine Geology*, *264*, 74–97, 2009

[25] Kim, J., F. Løvholt, D. Issler, and C. F. Forsberg, Landslide material control on tsunami genesis—the Storegga slide and tsunami (8,100 years bp), *Journal of Geophysical Research: Oceans*, 2019

[26] Najafi-Jilani, A., and B. Ataie-Ashtiani, Estimation of near-field characteristics of tsunami generation by submarine landslide, *Ocean Engineering*, *35*, 545–557, 2008

[27] Romano, A., M. Di Risio, M. G. Molfetta, G. Bellotti, D. Pasquali, P. Sammarco, L. Damiani, and P. De Girolamo, 3D physical modeling of tsunamis generated by submerged landslides at a conical island: The role of initial acceleration, *Coastal Engineering Proceedings*, *1*, 14, 2017

[28] Romano, A., J. L. Lara, G. Barajas, B. Di Paolo, G. Bellotti, M. Di Risio, I. J. Losada, and P. De Girolamo, Tsunamis generated by submerged landslides: numerical analysis of the near-field wave characteristics, *Journal of Geophysical Research: Oceans, n/a*, e2020JC016,157

[29] Perez del Postigo Prieto, N., A. Raby, C. Whittaker, and S. J. Boulton, Parametric study of tsunamis generated by earthquakes and landslides, *Journal of Marine Science and Engineering*, *7*, 154, 2019

[30] Heidarzadeh, M., T. Ishibe, O. Sandanbata, A. Muhari, and A. B. Wijanarto, Numerical modeling of the subaerial landslide source of the 22 December 2018 Anak Krakatoa volcanic

tsunami, Indonesia, *Ocean Engineering*, *195*, 106,733, 2020

[31] Heller, V., and W. H. Hager, Impulse product parameter in landslide generated impulse waves, *Journal of Waterway, Port, Coastal, and Ocean Engineering*, *136*, 145–155, 2010

[32] Mohammed, F., and H. M. Fritz, Physical modeling of tsunamis generated by three-dimensional deformable granular landslides, *Journal of Geophysical Research: Oceans (1978–2012)*, *117*, 2012

[33] Viroulet, S., A. Sauret, and O. Kimmoun, Tsunami generated by a granular collapse down a rough inclined plane, *EPL (Europhysics Letters)*, *105*, 34,004, 2014

[34] Lindstrøm, E. K., Waves generated by subaerial slides with various porosities, *Coastal Engineering*, *116*, 170–179, 2016

[35] Zitti, G., C. Ancey, M. Postacchini, and M. Brocchini, Impulse waves generated by snow avalanches: momentum and energy transfer to a water body, *Journal of Geophysical Research: Earth Surface*, *121*, 2399–2423, 2016

[36] Miller, G. S., W. A. Take, R. P. Mulligan, and S. McDougall, Tsunamis generated by long and thin granular landslides in a large flume, *Journal of Geophysical Research: Oceans*, *122*, 653–668, 2017

[37] Mulligan, R. P., and W. A. Take, On the transfer of momentum from a granular landslide to a water wave, *Coastal Engineering*, *125*, 16–22, 2017

[38] Tang, G., L. Lu, Y. Teng, Z. Zhang, and Z. Xie, Impulse waves generated by subaerial landslides of combined block mass and granular material, *Coastal Engineering*, *141*, 68–85, 2018

[39] Bullard, G., R. Mulligan, A. Carreira, and W. Take, Experimental analysis of tsunamis generated by the impact of landslides with high mobility, *Coastal Engineering*, *152*, 103,538, 2019

[40] Bellotti, G., C. Cecioni, and P. De Girolamo, Simulation of small-amplitude frequency-dispersive transient waves by means of the mild-slope equation, *Coastal Engineering*, *55*, 447–458, 2008

[41] Grilli, S. T., M. Shelby, O. Kimmoun, G. Dupont, D. Nicolsky, G. Ma, J. T. Kirby, and F. Shi, Modeling coastal tsunami hazard from submarine mass failures: effect of slide rheology, experimental validation, and case studies off the us east coast, *Natural Hazards*, *86*, 353–391, 2017

[42] Løvholt, F., C. B. Harbitz, and K. B. Haugen, A parametric study of tsunamis generated by submarine slides in the Ormen Lange/Storegga area off western Norway, in *Ormen Lange–an Integrated Study for Safe Field Development in the Storegga Submarine Area*, pp. 219–231, Elsevier, 2005

[43] Lynett, P., and P. L. F. Liu, A numerical study of the run-up generated by three-dimensional landslides, *Journal of Geophysical Research-Oceans*, *110*, 2005

[44] Ruffini, G., V. Heller, and R. Briganti, Numerical modelling of landslide-tsunami propagation in a wide range of idealised water body geometries, *Coastal Engineering*, p. 103518, 2019

[45] Si, P., H. Shi, and X. Yu, A general numerical model for surface waves generated by granular material intruding into a water body, *Coastal Engineering*, *142*, 42–51, 2018

[46] Watts, P., S. Grilli, J. Kirby, G. Fryer, and D. Tappin, Landslide tsunami case studies using a Boussinesq model and a fully nonlinear tsunami

generation model, *Natural Hazards And Earth System Sciences*, *3*, 391–402, 2003

[47] Whittaker, C., R. Nokes, H.-Y. Lo, P.-F. Liu, and M. Davidson, Physical and numerical modelling of tsunami generation by a moving obstacle at the bottom boundary, *Environmental Fluid Mechanics*, *17*, 929–958, 2017

[48] Abadie, S., D. Morichon, S. Grilli, and S. Glockner, Numerical simulation of waves generated by landslides using a multiple-fluid Navier–Stokes model, *Coastal Engineering*, *57*, 779–794, 2010

[49] Montagna, F., G. Bellotti, and M. Di Risio, 3D numerical modeling of landslide-generated tsunamis around a conical island, *Natural Hazards*, *58*, 591–608, 2011

[50] Ma, G., J. T. Kirby, T.-J. Hsu, and F. Shi, A two-layer granular landslide model for tsunami wave generation: Theory and computation, *Ocean Modelling*, *93*, 40–55, 2015

[51] Ma, G., F. Shi, and J. T. Kirby, Shock-capturing non-hydrostatic model for fully dispersive surface wave processes, *Ocean Modelling*, *43*, 22–35, 2012

[52] Heller, V., M. Bruggemann, J. Spinneken, and B. D. Rogers, Composite modelling of subaerial landslide–tsunamis in different water body geometries and novel insight into slide and wave kinematics, *Coastal Engineering*, *109*, 20–41, 2016

[53] Crespo, A. J., J. M. Domnguez, B. D. Rogers, M. Gómez-Gesteira, S. Longshaw, R. Canelas, R. Vacondio, A. Barreiro, and O. Garca-Feal, Dualsphysics: Open-source parallel cfd solver based on smoothed particle hydrodynamics (sph), *Computer Physics Communications*, *187*, 204–216, 2015

[54] Shi, C., Y. An, Q. Wu, Q. Liu, and Z. Cao, Numerical simulation of landslide-generated waves using a soil–water coupling smoothed particle hydrodynamics model, *Advances in Water Resources*, *92*, 130–141, 2016

[55] Fritz, H., W. Hager, and H.-E. Minor, Lituya bay case: Rockslide impact and wave run-up, *Science of Tsunami Hazards*, *19*, 3–22, 2001

[56] Jasak, H., Error analysis and estimation for the finite volume method with applications to fluid flows, *(Ph.D thesis) Imperial College London (University of London)*, 1996

[57] Higuera, P., J. L. Lara, and I. J. Losada, Realistic wave generation and active wave absorption for Navier–Stokes models: Application to OpenFOAM®, *Coastal Engineering*, *71*, 102–118, 2013

[58] Higuera, P., J. L. Lara, and I. J. Losada, Simulating coastal engineering processes with OpenFOAM®, *Coastal Engineering*, *71*, 119–134, 2013

[59] Higuera, P., J. L. Lara, and I. J. Losada, Three-dimensional interaction of waves and porous coastal structures using OpenFOAM®. Part I: formulation and validation, *Coastal Engineering*, *83*, 243–258, 2014

[60] Higuera, P., J. L. Lara, and I. J. Losada, Three-dimensional interaction of waves and porous coastal structures using OpenFOAM®. Part II: Application, *Coastal Engineering*, *83*, 259–270, 2014

[61] Kim, G.-B., W. Cheng, R. C. Sunny, J. J. Horrillo, B. C. McFall, F. Mohammed, H. M. Fritz, J. Beget, and Z. Kowalik, Three dimensional landslide generated tsunamis: Numerical and physical model comparisons, *Landslides*, pp. 1–17, 2019

[62] Clous, L., and S. Abadie, Simulation of energy transfers in waves generated by granular slides, *Landslides*, pp. 1–17, 2019

[63] MiDi, G., On dense granular flows, *The European Physical Journal E*, *14*, 341–365, 2004

[64] Mulligan, R. P., A. Franci, M. A. Celigueta, and W. A. Take, Simulations of landslide wave generation and propagation using the particle finite element method, *Journal of Geophysical Research: Oceans*, *125*, e2019JC015,873, 2020

[65] Chen, F., V. Heller, and R. Briganti, Numerical modelling of tsunamis generated by iceberg calving validated with large-scale laboratory experiments, *Advances in Water Resources*, p. 103647, 2020

[66] Heller, V., T. Attili, F. Chen, M. Brühl, R. Gabl, X. Chen, G. Wolters, and H. Fuchs, Large-scale experiments of tsunamis generated by iceberg calving, in *38th International Association for Hydro-Environmental Engineering and Research World Congress (IAHR 2019)*, 2019

[67] Jasak, H., D. Rigler, and Ž. Tuković, Design and implementation of Immersed Boundary Method with discrete forcing approach for boundary conditions, in *11th World Congress on Computational Mechanics, WCCM 2014, 5th European Conference on Computational Mechanics, ECCM 2014 and 6th European Conference on Computational Fluid Dynamics, ECFD 2014*, pp. 5319–5332, 2014

[68] del Jesus, M., J. L. Lara, and I. J. Losada, Three-dimensional interaction of waves and porous coastal structures: Part I: Numerical model formulation, *Coastal Engineering*, *64*, 57–72, 2012

[69] Lara, J. L., M. del Jesus, and I. J. Losada, Three-dimensional interaction of waves and porous coastal structures: Part II: Experimental validation, *Coastal Engineering*, *64*, 26–46, 2012

[70] Losada, I. J., J. L. Lara, and M. del Jesus, Modeling the interaction of water waves with porous coastal structures, *Journal of Waterway, Port, Coastal, and Ocean Engineering*, *142*, 03116,003, 2016

Chapter 9

Hydrodynamics of Regular Breaking Wave

Diana De Padova and Michele Mossa

Abstract

Turbulence and undertow currents play an important role in surf-zone mixing and transport processes; therefore, their study is fundamental for the understanding of nearshore dynamics and the related planning and management of coastal engineering activities. Pioneering studies qualitatively described the features of breakers in the outer region of the surf zone. More detailed information on the velocity field under spilling and plunging breakers can be found in experimental works, where single-point measurement techniques, such as Hot Wire Anemometry and Laser Doppler Anemometry (LDA), were used to provide maps of the flow field in a time-averaged or ensemble-averaged sense. Moreover, the advent of non-intrusive measuring techniques, such as Particle Image Velocimetry (PIV) provided accurate and detailed instantaneous spatial maps of the flow field. However, by correlating spatial gradients of the measured velocity components, the instantaneous vorticity maps could be deduced. Moreover, the difficulties of measuring velocity due to the existence of air bubbles entrained by the plunging jet have hindered many experimental studies on wave breaking encouraging the development of numerical model as useful tool to assisting in the interpretation and even the discovery of new phenomena. Therefore, the development of an WCSPH method using the RANS equations coupled with a two-equation k–ε model for turbulent stresses has been employed to study of the turbulence and vorticity distributions in in the breaking region observing that these two aspects greatly influence many coastal processes, such as undertow currents, sediment transport and action on maritime structures.

Keywords: regular breaking waves, shear stress, turbulence, kinetic energy, vorticity, physical modeling, numerical modeling

1. Introduction

Wave breaking is one of the most important process for coastal engineers since it greatly influences both the transport processes and the magnitudes of the forces on coastal structures [1, 2]. Wave breaking in the surf zone drives complicated turbulent structures and for this reason breaking is possibly the most difficult wave phenomenon to describe mathematically [3–5].

Pioneering experimental studies were carried out by [6–11], who described the velocity field under plunging breakers in the outer region of the surf zone; more recently by [12–17]. As observed experimentally by [18], during the pre-breaking stages, the maximum turbulence intensity appears at the core of the main vortex and decreases as the vortex moves downstream. Additional turbulence is then generated near the free surface during the breaking process.

Moreover, the difficulties of measuring velocity due to the existence of air bubbles entrained by the plunging jet have hindered many experimental studies on wave breaking encouraging the development of numerical model as useful tool to assisting in the interpretation and even the discovery of new phenomena. One of the great advantages of the numerical models is their ability to disclose the evolutions of undertow currents and turbulence quantities in the spatial and temporal domains, which are too expensive to be investigated by experiments. Therefore, the main emphasis for research is placed on the application and development of numerical methods. Furthermore, for consistent and accurate results, it is essential to calibrate the numerical models with experimental data.

The numerical models can be classified as Eulerian or Lagrangian method. In Eulerian method, the space is discretized into a grid or mesh and the unknown values are defined at the fix points, while a Lagrangian method tracks the pathway of each moving mass point. The Eulerian methods such as the finite difference methods (FDM), finite volume methods (FVM) and the finite element methods (FEM) have been widely applied in many fields of engineering because are very useful to solve differential or partial differential equations (PDEs) that govern the concerned physical phenomena [19–23]. Despite the great success, grid based numerical methods suffer from difficulties in some aspects such as the use of grid/mesh makes the treatment of discontinuities (e.g., wave breaking, cracking and contact/separation) difficult because the path of discontinuities may not coincide with the mesh lines.

Therefore, during the last years, research has been focused on Lagrangian techniques such as Discrete Element Method (DEM) [24], Smoothed Particle Hydrodynamics (SPH) [25], Immersed Particle Method (IPM) [26, 27]. The development and applications of the major existing Lagrangian methods have been addressed in some review articles such as [28–30]. In general, the Lagrangian methods provide accurate and stable numerical solutions for integral equations or partial differential equations (PDEs) with all kinds of possible boundary conditions using a set of arbitrarily distributed nodes or particles. During the last years, Smoothed Particle Hydrodynamics (SPH) has become a very powerful method for CFD problems governed by the Navier–Stokes equations such as fluid-dynamic problems with highly non-linear deformation [31–37]; multi-phase flows for coastal and other hydraulic applications with air-water mixture sand sediment scouring [38–42]; oscillating jets inducing breaking waves [43] and nonbuoyant jets in a wave environment [44–46]; fluid/structure/soil-interaction [47–49]; hydraulic jumps [50–53]; multi-phase flows and oil spill [54–55].

The present chapter is organized as follows. First, an WCSPH method is developed using the RANS equations and a two-equation k–ε model is formulated using the particle approach. Then the numerical model is employed to reproduce breaking in spilling and plunging waves in a sloped wave channel. The experimental data by [14] are used to check the model results. This reveals the importance of experimental data in these studies. The present chapter is aimed to describe some recent results obtained within the frame of numerical and experimental analyses of wave breaking. The new insight is the investigation of the ability of WCSPH with a k–ε turbulence closure model to disclose the turbulence dispersion and the temporal and spatial evolutions of turbulence quantities in different types of breakers.

2. Mathematical formulation

A Lagrangian numerical model is developed to solve free surface turbulent flows. The flow field is governed by the Reynolds Averaged Navier–Stokes (RANS) equations and the k–ε turbulence equations [56]. These equations are solved by the WCSPH method in which an artificial compressibility is introduced to solve explicitly in time the equations of motion of an incompressible fluid.

Using the SPH approach, the fluid flow domain is initially represented by a finite number of particles. These particles can be viewed as moving numerical nodes, which move according to the governing equations and boundary conditions. Each discrete point is associated to an elementary fluid volume (or particle) i, which has position xi and constant mass mi.

To find the value of $\boldsymbol{a}(\overline{\boldsymbol{x}}, t)$ at a generic point $\overline{\boldsymbol{x}}$ an interpolation is applied from the nodal values ai(t) through a kernel function $W = (\boldsymbol{x} - \overline{\boldsymbol{x}}, \eta)$ as follows:

$$\boldsymbol{a}(\bar{\boldsymbol{x}}, t) \approx < \boldsymbol{a}(\bar{\boldsymbol{x}}, t) > = \sum_{j=1}^{N} \frac{m_j}{\rho_j} \boldsymbol{a}(\boldsymbol{x}_j, t) W(\boldsymbol{x}_j - \bar{\boldsymbol{x}}, \eta) \tag{1}$$

where ρ is the fluid density, and the summation is extended to all the N particles located inside the sphere of radius 2η centered on $\bar{\boldsymbol{x}}$. The *kernel* function is continuous, non-zero only inside a sphere $\boldsymbol{x} - \overline{\boldsymbol{x}} < 2\eta$ and tends to the Dirac delta function when η (defined as the smoothing length) tends to zero. There are different available kernel functions and the kernel operations can be inaccurate for cases where the particle distribution is non-uniform or the support for the interpolations is incomplete [55]; Quinlan et al., [57], Randles and Libersky [58] and Bonet and Lok [59] introduced a Kernel correction which ensure at least first-order consistency; however the corrected kernel is non-symmetric which leads to non-conservative interpolations.

Dehnen and Aly [60] showed that the Wendland *kernel* function [61] is more computationally convenient than the B-spline function, allowing better numerical convergence; Liu and Liu [62] showed that the quintic-spline function [63] is more effective in interpolating the second-order derivatives. The SPH computations discussed in the present paper were based on the cubic-spline kernel function proposed by [64] that is more effective in the simulation of several different hydraulic flows [65, 66].

The advantage of SPH approach is that differential operator applied to $\boldsymbol{a}(\overline{\boldsymbol{x}}, t)$ can be approximated by making use of the gradient of the kernel function. For instance, the divergence $\nabla \cdot \boldsymbol{a}(\overline{\boldsymbol{x}}, t)$ can be approximated by:

$$\nabla \cdot \boldsymbol{a}(\bar{\boldsymbol{x}}, t) \approx < \nabla \cdot \boldsymbol{a}(\bar{\boldsymbol{x}}, t) > = \sum_{j=1}^{N} \frac{m_j}{\rho_j} \left[\boldsymbol{a}(\bar{\boldsymbol{x}}, t) - \boldsymbol{a}(\boldsymbol{x}_j, t)\right] \nabla W(\boldsymbol{x}_j - \bar{\boldsymbol{x}}, \eta) \tag{2}$$

For further details on the different methods for SPH approximations of all the vector operators, the reader can see [67, 68]. In a Lagrangian frame, the Reynolds-averaged Navier–Stokes (RANS) equations and the k–ε turbulence equations take the following form

$$\begin{cases} \dfrac{D\rho}{Dt} + \rho \nabla \cdot \boldsymbol{v} = 0 \\[2ex] \dfrac{D\boldsymbol{v}}{Dt} = \dfrac{1}{\rho} \nabla p + \dfrac{1}{\rho} \nabla \bullet T + \boldsymbol{g} \\[2ex] p - p_0 = c^2(\rho - \rho_0); T = \mu_T S \end{cases} \tag{3}$$

where $\boldsymbol{v} = (u, v)$ is the velocity vector, p is pressure, $\mathbf{g}$ is the gravity acceleration vector, T is the turbulent shear stress tensor, c is the speed of sound in the weakly compressible fluid, μ_T is the dynamic eddy viscosity, S is rate-of-strain tensor and the subscript 0 denotes a reference state for pressure computation.

The RANS equations (3) in the SPH semi-discrete form become

$$\begin{cases} \left\langle \frac{D\rho_i}{Dt} \right\rangle = \sum_j m_j (v_i - v_j) \nabla W_{ij} \\ \left\langle \frac{D\boldsymbol{v_i}}{Dt} \right\rangle = -\sum_j m_j \left(\frac{p_i}{\rho_i^2} + \frac{p_j}{\rho_j^2} \right) \nabla W_{ij} + \sum_j \frac{m_j}{\rho_j} (T_i - T_j) \bullet \nabla W_{ij} + \boldsymbol{g} \\ p_i - p_0 = c_i^2 (\rho_i - \rho_0);\ T_i = \mu_{T_i} S_i \end{cases} \tag{4}$$

where W_{ij} is a shorthand notation for $W(\boldsymbol{x}_i - \boldsymbol{x}_j, \eta)$, renormalized through a procedure which enforces consistency on the first derivatives to the 1st order [69], leading to a 2nd order accurate discretization scheme in space. The semi-discretized system (4) is then integrated in time by a 2nd order two-step XSPH explicit algorithm [70].

The momentum equation is then solved to yield an intermediate velocity field $\hat{\boldsymbol{v}}$, which is then corrected through a smoothing procedure based on the values of the neighboring fluid particles.

$$\boldsymbol{v}_i = (1 - \phi_v)\hat{\boldsymbol{v}}_i + \phi_v \frac{\sum_j \frac{m_j}{\rho_j} \hat{\boldsymbol{v}}_j W_{ij}}{\sum_j \frac{m_j}{\rho_j} W_{ij}} \tag{5}$$

using a velocity smoothing coefficient φ_v. The corrected velocity value is then used to update the particle position and to solve the continuity equation. The new density values are finally used to compute pressure, according to the equation of state.

A pressure smoothing procedure is also applied to the difference between the local and the hydrostatic pressure values [71] in order to reduce the numerical noise in pressure evaluation which is present, in particular in WCSPH, owing to high frequency acoustic signals [72]. The present method is applied only to the difference between the intermediate pressure field $\hat{p}$ and the hydrostatic pressure gradient to ensure the conservation of total volume of the particle system for long time simulations [73].

The eddy viscosity coefficient $\mu_T = c_\mu \frac{k^2}{\varepsilon}$ in Eq. (4) was evaluated through a k-ε model by [74]:

$$\begin{cases} \frac{Dk_i}{Dt} = P_{k_i} + \frac{1}{\sigma_k} \sum_j m_j \frac{\nu_{T_i} + \nu_{T_j}}{\rho_i + \rho_j} \frac{k_i - k_j}{r_{ij}^2 + 0.01h^2} \boldsymbol{r}_{ij} \cdot \nabla W_{ij} - \varepsilon_i \\ \frac{D\varepsilon_i}{Dt} = \frac{1}{\sigma_\varepsilon} \sum_j m_j \frac{\nu_{T_i} + \nu_{T_j}}{\rho_i + \rho_j} \frac{\varepsilon_i - \varepsilon_j}{r_{ij}^2 + 0.01h^2} \boldsymbol{r}_{ij} \cdot \nabla W_{ij} + +C_{\varepsilon_1} \frac{\varepsilon_i}{k_i} P_{k_i} + C_{\varepsilon_2} \frac{\varepsilon_i}{k_i} \sum_j \frac{m_j}{\rho_j} \varepsilon_j W_{ij} \end{cases} \tag{6}$$

where k_i is the turbulent kinetic energy per unit mass, ε is the dissipation rate of turbulent kinetic energy, P_k is the production of turbulent kinetic energy depending on the local rate of deformation and ν_T is the eddy viscosity and $\boldsymbol{r}_{ij} = \boldsymbol{x}_i - \boldsymbol{x}_j$. There are several empirical coefficients in the k–ε turbulent closure model. In this paper the set of constant values recommended by [74], i.e., $c_\mu = 0.09$, $\sigma_k = 1$, $\sigma_\varepsilon = 1.3$, $C_{\varepsilon_1} = 1.44$ and $C_{\varepsilon_2} = 1.92$, is adopted.

According to Eq. (6) both the production term and dissipation term for ε become singular when k approaches zero. Furthermore, no turbulence energy can be produced if there is no turbulent kinetic energy initially. Thus, it is necessary to "seed" a small amount of k in both the initial condition and inflow boundary condition. In this paper the initial seeding of turbulent kinetic energy recommended by [75] is adopted.

3. Validation and application

3.1 Experimental set up

The results obtained from the numerical model outlined in the previous section have been validated against extensive experimental data [14], and then used to obtain further insight in the physics of the flow here analyzed.

The detailed experimental setup has been given in De Serio and Mossa [14]. Here only some important parameters are summarized.

Experiment was carried out in the wave flume 45 m long and 1 m wide of the *Department of Civil, Environmental, Land, Building Engineering and Chemistry* (DICATECh) of the *Polytechnic University of Bari* (Italy). A beach with constant slope of 1/20 is connected to a region with constant water depth of h = 0.7 m. The wave generating system is a piston-type one, with paddles producing the desired wave by providing a translation of the water mass, according to the proper input signal. The instantaneous Eulerian velocities were acquired by a backscatter, two-component, four beam Laser Doppler Anemometer (LDA) system and a Dantec LDA signal processor (58 N40 FVA Enhanced) based on the covariance technique. The wave elevations were measured with a resistance probe placed in the transversal section of the channel crossing the laser measuring volume.

Figure 1a–f show the different parts of the complex experimental apparatus, which comprises the LDA system, the resistance wave gauge system and the wavemaker system. Further details about the experimental tests can be found in [14].

A sketch view of the experimental setup is shown in **Figure 2**.

Table 1 shows the main parameters of the examined waves listed for each experiment, such as the offshore wave height H_O, the wave period T and the deepwater wavelength L_O, estimated in section 76, where the bottom is flat and the mean water depth h is equal to 0.70 m. In the experiments, the regular wave had a height H_O = 11 cm and a period T = 2.0 s for the spilling breaker case (T1), while H_O = 6.5 cm and T = 4.0 s were used to generate a plunging breaker (T2). **Table 1** shows also the Irribarren number ξ_O, computed for the two tests from the following equation

$$\xi_0 = \frac{\tan\beta}{\sqrt{\frac{H_0}{L_0}}} \tag{7}$$

in which β is the bottom slope angle.

Water surface elevations and velocities were measured at six different sections along the longitudinal axis of symmetry of the wave channel named 76, 55, 49, 48, 47, 46 and 45 (see **Table 2**). Specifically, for all two tests, section 48 was in the pre-breaking region, section 47 was where the incipient breaking occurred, while in sections 46 and 45, the wave re-arranged into a bore.

3.2 Numerical model setup and validation

The WCSPH method coupled with a k–ε turbulence model has been employed to reproduce the above experiment. The computational domain has been reduced to be 20.0 m long so as to save computing expenses (**Figure 3**).

Figure 1.
Experimental apparatus: (a) LDA probe; (b) DANTEC FVA signal processor and process computer; (c) laser coherent Innova and Dantec 2D fiber flow optics; (d) process computer with a AD/DA board for the wavemaker control; (e) a part of the wave channel; (f) the wavemaker.

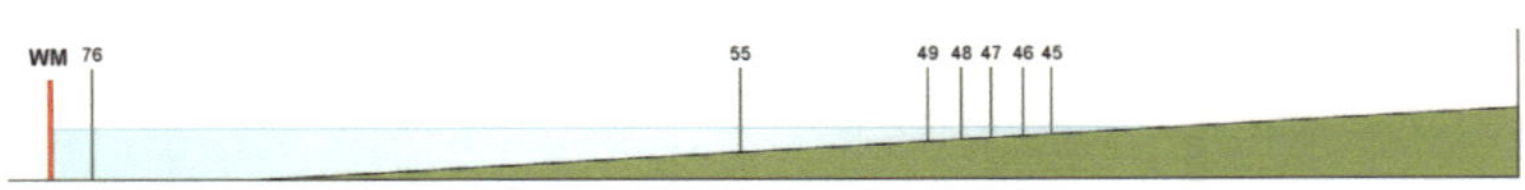

Figure 2.
Sketch view of experimental setup.

	H_0 (cm)	T (s)	L_0 (m)	d (m)	ξ_0	Breaking type
T1	11	2	4.62	0.70	0.37	Spilling/plunging
T2	6.5	4	10.12	0.70	0.74	Plunging

Table 1.
Experimental parameters of the analyzed regular waves.

Investigated section	Distance from paddles (m)	*d* (cm)
Section 76	10.56	70.0
Section 55	19.80	31.0
Section 49	22.44	16.5
Section 48	22.88	14.0
Section 47	23.32	11.3
Section 45	24.20	8.5

Table 2.
Location of measurement sections.

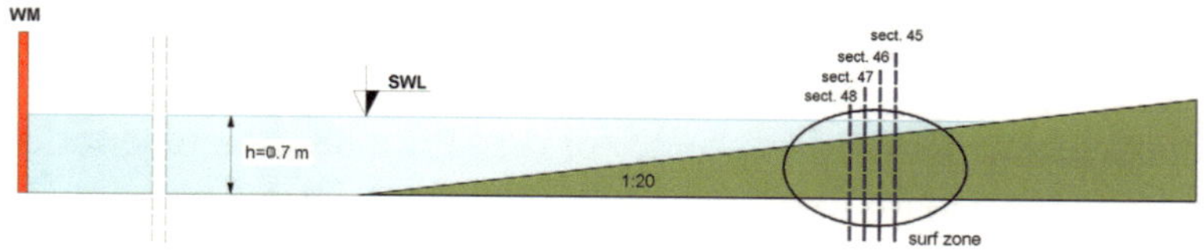

Figure 3.
Sketch of the computational domain wave channel with location of the seven investigated sections, used to calibrate the numerical model.

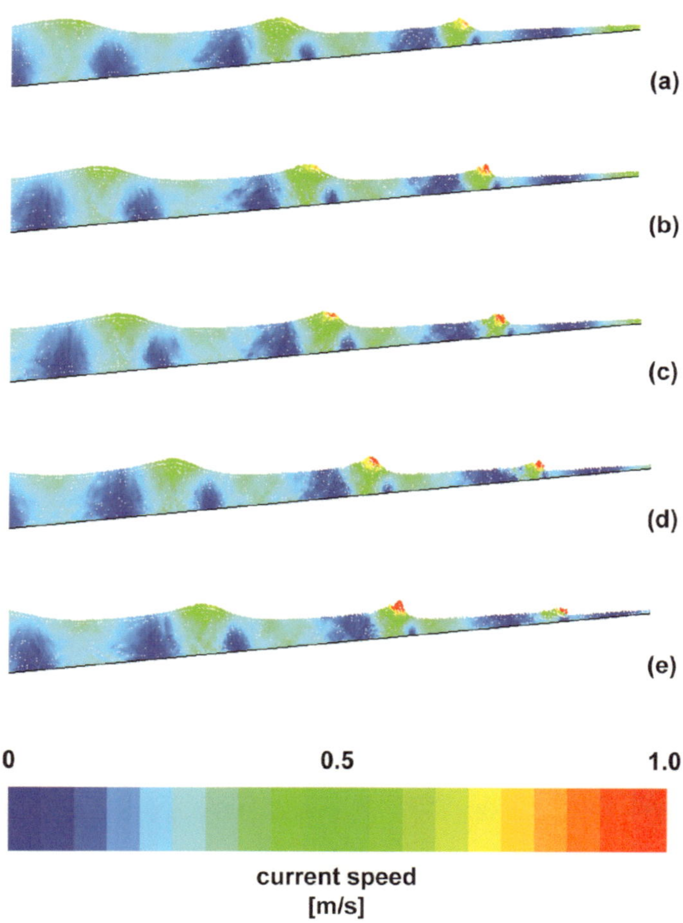

Figure 4.
Instantaneous SPH velocity field in the SPH simulation of spilling wave (T1): (a) before; (b)–(c) during and (d)–(e) after breaking.

The adopted offshore boundary condition guarantees a regular development of the wave train before the sloping section of the channel and, therefore, does not influence the quality of the numerical solution, as shown by [32].

For both the two tests, the offshore boundary condition has been treated as dynamic boundary condition modeled by a numerical wave paddle also composed of ghost particles whose motion has been forced to obtain the frequency and amplitude of the wave paddle needed to generate the desired sinusoidal wave [76]. The initial water depth was set equal to 0.70 m. In the present simulations, the initial particle spacing Σ = 0.022 m, the value of η/Σ = 1.5 and φ = 0.01, recommended by De Padova et al. [32], have been adopted.

The instantaneous SPH particle distribution and velocity magnitude snapshots of the breaking wave are shown in **Figures 4a–e** and **5a–e**, respectively, for the spilling and plunging breakers. These results show that the general features of wave breaking, collapsing and a turbulent bore propagating have been well captured by the SPH computations.

In order to further verify the accuracy of the SPH model the time series of wave elevations, horizontal and vertical velocities at the investigated sections (**Figure 3**)

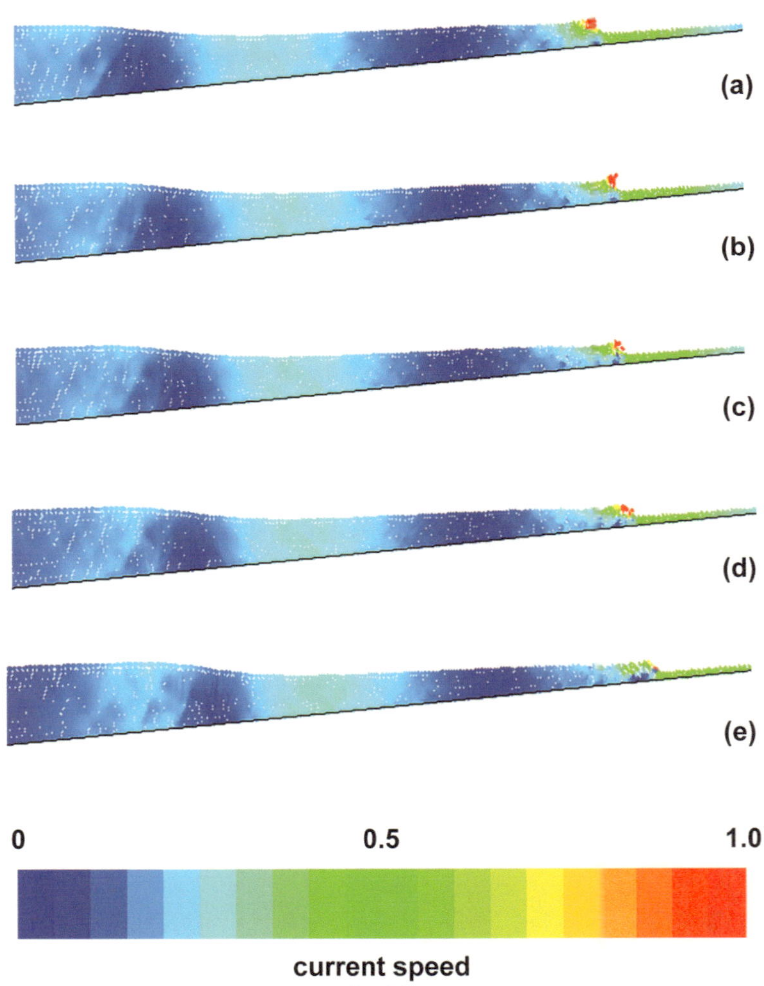

Figure 5.
Instantaneous SPH velocity field in the SPH simulation of plunging wave (T2): (a) before; (b)–(c) during and (d)–(e) after breaking.

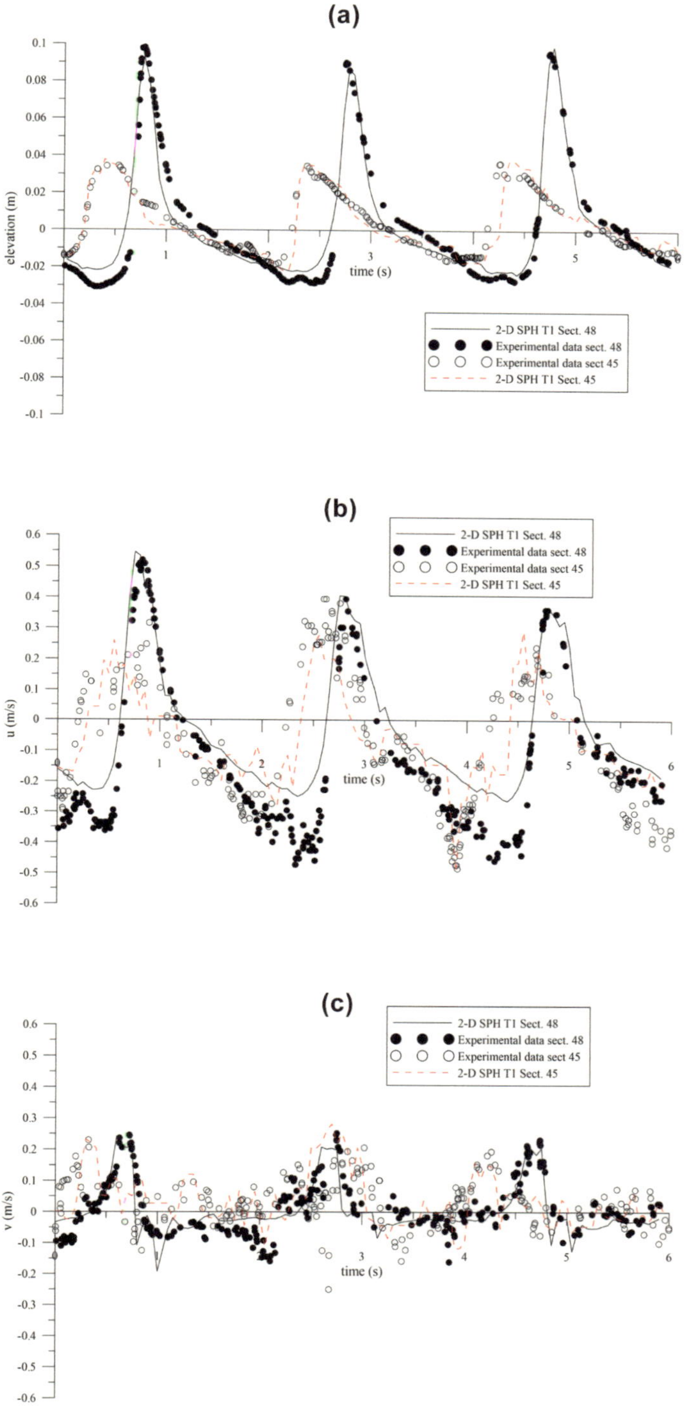

Figure 6.
Instantaneous computed and measured (a) wave elevations, (b) horizontal and (c) vertical velocities in section 48 and section 45 for T1.

have been compared with the experimental data of De Serio and Mossa [14].
As an example, in **Figure 6a–c** both laboratory and numerical wave surface elevations, and velocities at vertical sections 48 and 45 are plotted for T1, referring to the point located at 1 cm from the bottom. The agreement between the calibrated numerical results and the laboratory measurements is fairly good.

4. Results and discussion

One of the great advantages of the numerical models is their ability to show the evolutions of vorticity and turbulence quantities in the spatial and temporal domains, which are too expensive to be investigated by experiments. Using the SPH computational results, the turbulent kinetic energy distributions are shown in **Figures 7a–e** and **8a–e**, respectively, for the spilling (T1) and plunging (T2) waves. For both breakers, the turbulence quantity has the largest values near the free surface and decreases into the water column. However, the results highlight that there exist fundamental differences in the dynamics of turbulence between the spilling and plunging breakers, which can be related to the processes of wave breaking production.

For the spilling wave (T1), higher turbulence levels are mainly concentrated in the breaking wave front and the highest turbulence level appears in the roller region (**Figure 7d**). After the breaking, as the wave propagates forward, the turbulence kinetic energy decreases (**Figure 7e**). Instead, the turbulence levels increase rapidly after the wave breaking for the plunging case (T2) as shown in **Figure 8c–e**.

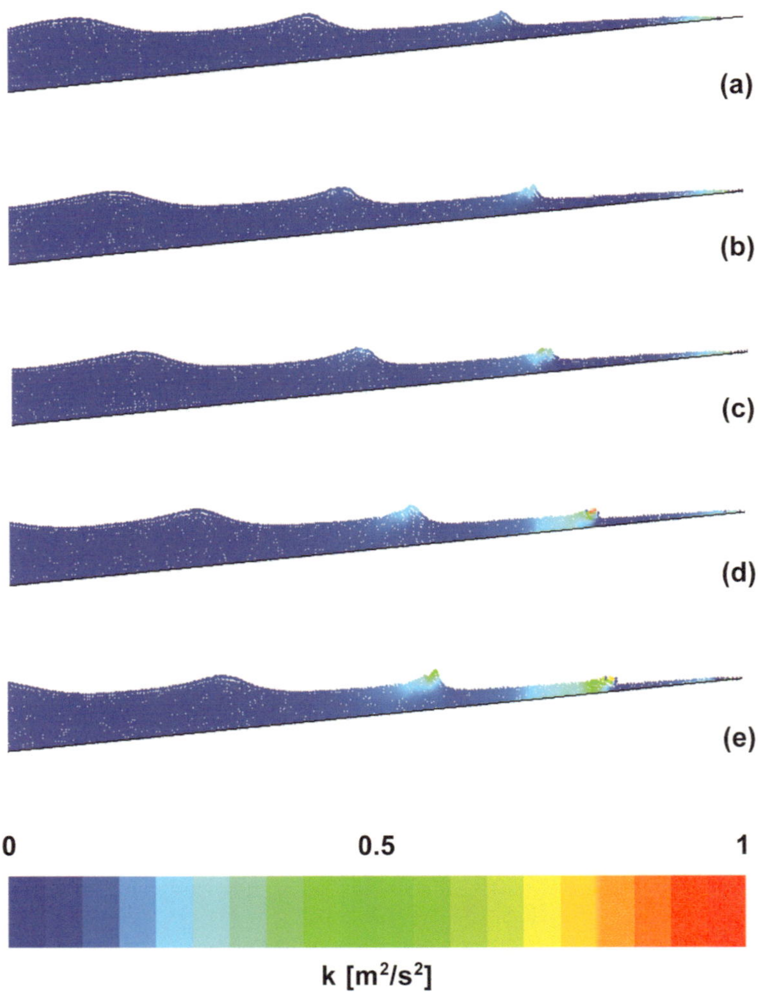

Figure 7.
Instantaneous turbulence intensity distributions in the SPH simulation of spilling wave (T1): (a) before; (b)–(c) during and (d)–(e) after breaking.

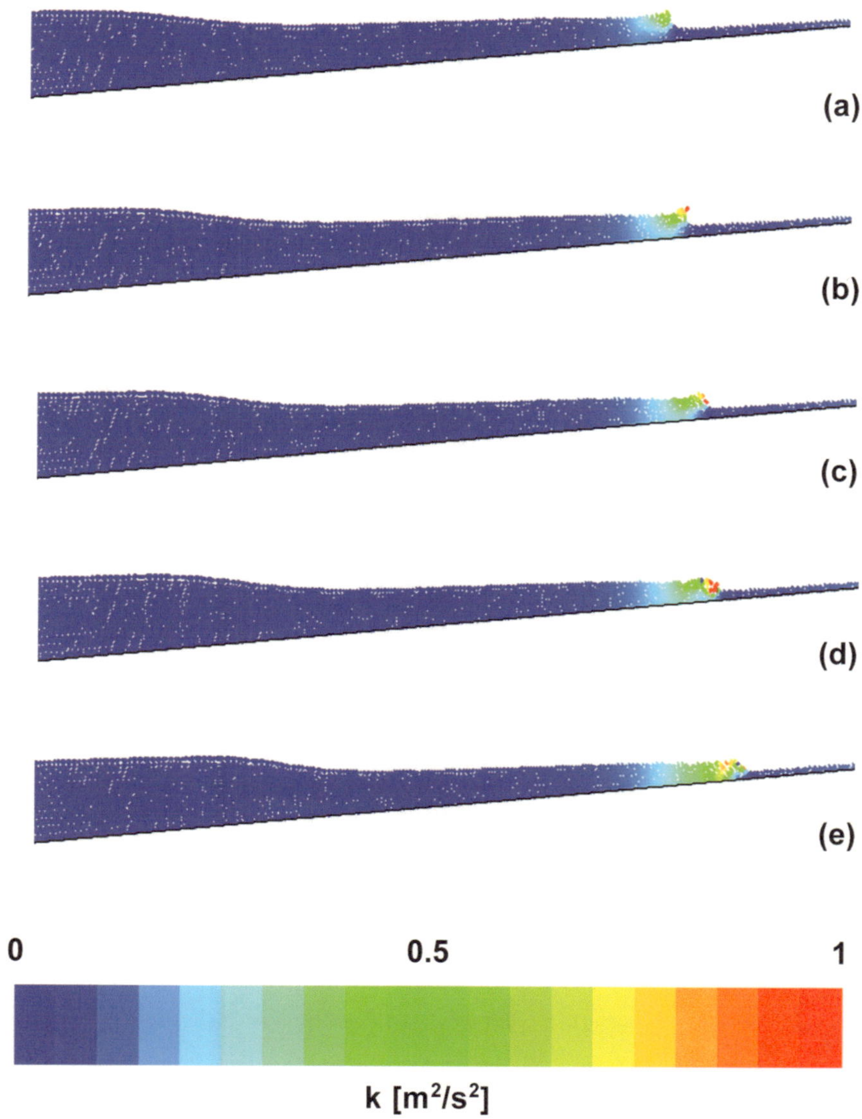

Figure 8.
Instantaneous turbulence intensity distributions in the SPH simulation of plunging wave (T2): (a) before; (b)–(c) during and (d)–(e) after breaking.

The maximum turbulence level is generated as the plunging jet touches down on the wave trough (**Figure 8d**) in sections 46–45 (**Figure 2**); After the breaking, the roller continues to spread downwards and therefore high turbulence levels are generated beneath the free surface after breaking (**Figure 8e**).

Using the SPH computational results, the vorticity maps are shown in **Figures 9a–e** and **10a–e**, respectively, for the spilling and plunging waves. Vorticity is defined as

$$\omega = \left(\frac{\partial u}{\partial z}\right) - \left(\frac{\partial v}{\partial x}\right) \tag{8}$$

and is computed using instantaneous values of the horizontal and vertical velocity.

As noted by several authors [77, 78], for both breakers (T1 and T2), when the breaking begins, positive vorticity occupies the whole region of the surface roller

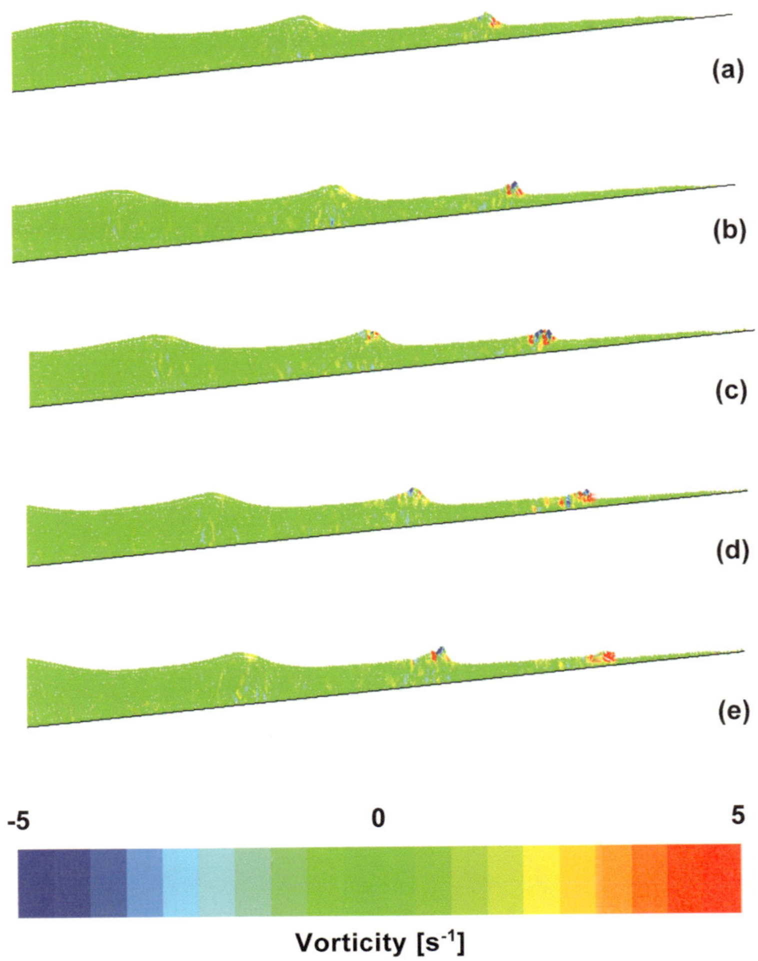

Figure 9.
Instantaneous values of ω distributions in the SPH simulation of spilling wave (T1): (a) before; (b)–(c) during and (d)–(e) after breaking.

and spreads out over the whole water column. However, the vorticity levels increase rapidly after the wave breaking for the plunging case (T2) due to the strong impingement of the jet on the forward trough, inducing a propagation of the positive vorticity towards the bottom (**Figure 10c–e**).

Moreover, the results highlight that there exist differences in the dynamics of vorticity between the spilling and plunging breakers. In fact, only during spilling formation (T1), small structures of negative vorticity are generated, instead when the plunging breaker (T2) occurs the fluid is relatively free of negative vorticity regions.

Figures 11 and **12** show the comparison between the instantaneous map of vorticity and of the surface parallel convective acceleration for the spilling and plunging waves (T1 and T2) when the breaking begins at time step of **Figures 9b** and **10b**, respectively. The surface parallel convective acceleration here has been computed following [79]. As noted by Dabiry and Gharib [80], for both breakers (T1 and T2), a flow deceleration (**Figures 11b** and **12b**) occurs in the same location where peaks of positive vorticity appear (**Figures 11a** and **12a**). Therefore, the present results confirm the findings by Dabiri and Gharib [80] that the vorticity is convected due to the sharp velocity gradient of the fluid near the free surface with respect to the fluid below.

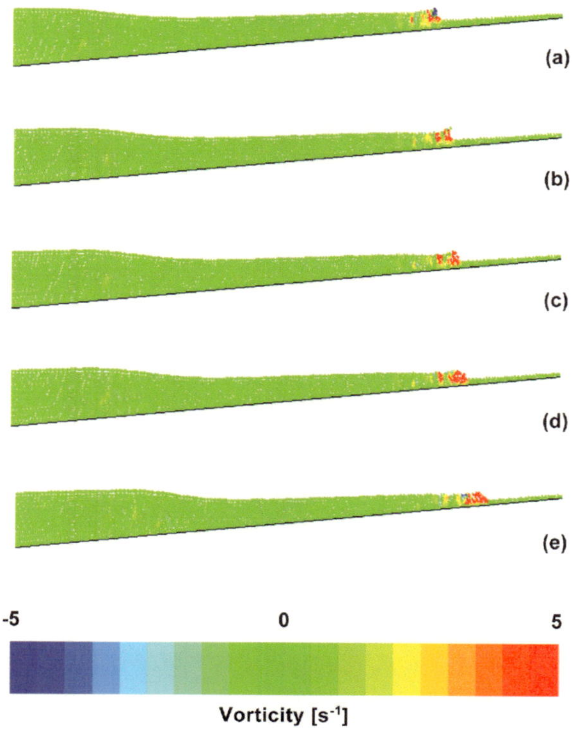

Figure 10.
Instantaneous values of ω distributions in the SPH simulation of spilling wave (T2): (a) before; (b)–(c) during and (d)–(e) after breaking.

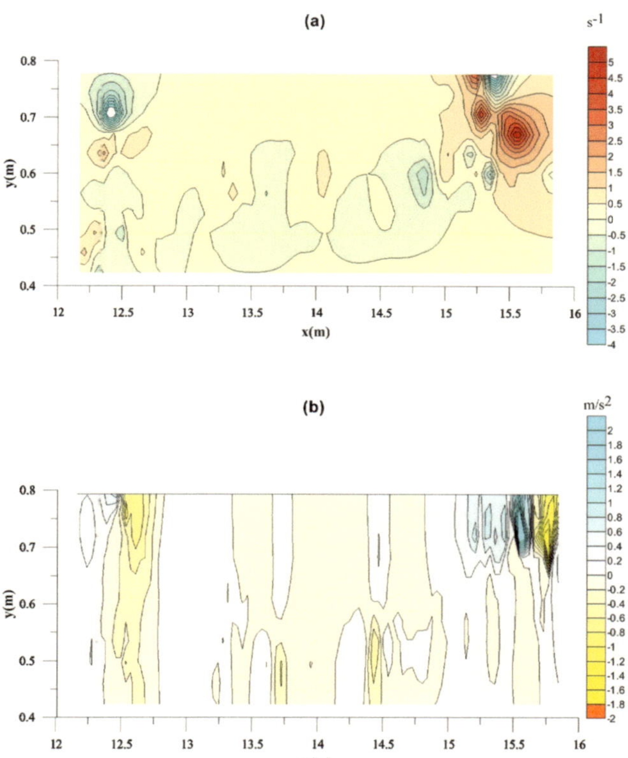

Figure 11.
SPH simulation of spilling wave (T1): (a) Vorticity map and (b) surface-parallel convective acceleration map.

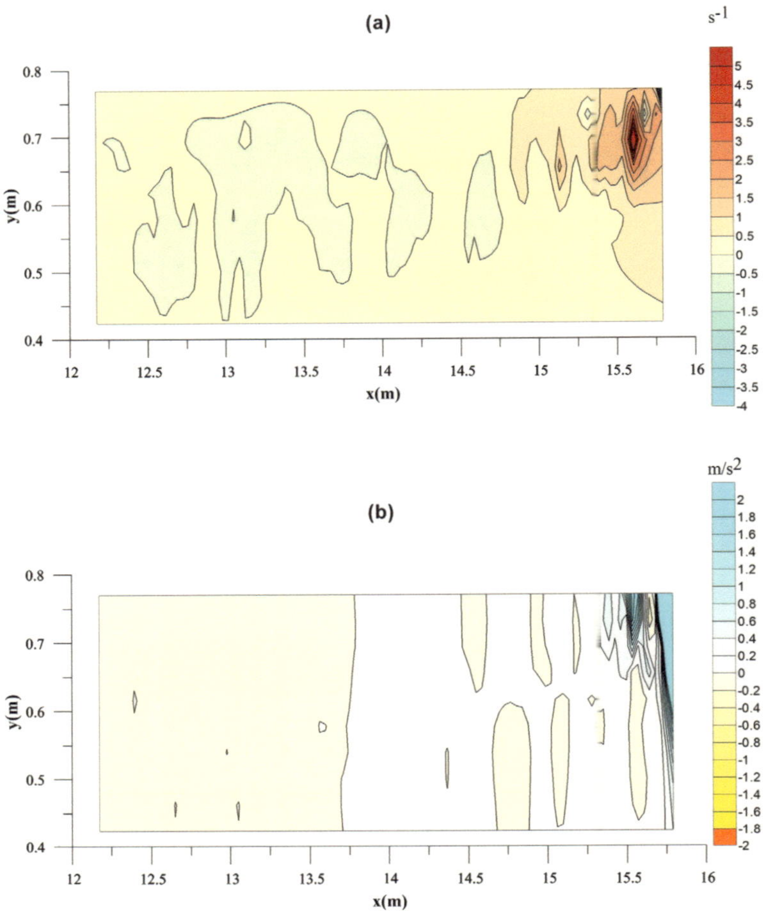

Figure 12.
SPH simulation of plunging wave (T2): (a) Vorticity map and (b) surface-parallel convective acceleration map.

5. Conclusions

In the present chapter a WCSPH method has been developed using the RANS equations and a two-equation k–ε model has been formulated using the particle approach. Then the numerical model has been employed to reproduce breaking in spilling and plunging waves in a sloped wave channel. The experimental data by [14] have been used to check the model results. Finally, we have fully exploited the advantages of numerical modeling to disclose fundamental differences between the different types of breakers by investigating the temporal and spatial evolutions of turbulence quantities and vorticity field.

For the spilling wave (T1), during the pre-breaking and breaking stages, the maximum turbulence intensity has been generated near the free surface and decreases as the vortex moves downstream. Instead, the turbulence levels increased rapidly after the wave breaking for the plunging case (T2). In fact, the maximum turbulence level was generated as the plunging jet touches down on the wave trough and after the breaking, the roller continued to spread downwards and therefore high turbulence levels were generated beneath the free surface after breaking.

For both breakers (T1 and T2), analyzing the instantaneous vorticity distributions, when the breaking begins, positive vorticity has occupied the whole region of the surface roller and has spread out over the whole water column. However, only during spilling formation (T1), small structures of negative vorticity have been generated, instead when the plunging breaker (T2) occurs the fluid was relatively free of negative vorticity regions.

Furthermore, comparing the instantaneous map of vorticity and of the surface parallel convective acceleration for the spilling and plunging waves (T1 and T2), the present results confirmed the findings by Dabiri and Gharib [80] that the vorticity was convected due to the sharp velocity gradient of the fluid near the free surface with respect to the fluid below.

Author details

Diana De Padova and Michele Mossa*
Department of Civil, Environmental, Land, Building Engineering and Chemistry (DICATECh), Polytechnic University of Bari, Bari, Italy

*Address all correspondence to: michele.mossa@poliba.it

References

[1] D. B. Clark, S. Elgar, and B. Raubenheimer. Vorticity generation by short-crested wave breaking. Geophysical Research Letters, 39(24), 2012.

[2] Z. Wei, C. Li, R. A. Dalrymple, M. Derakhti, and J. Katz. Chaos in breaking waves. Coastal Engineering, 140:272–291, 2018.

[3] Nadaoka K, Hino M, Koyano Y. Structure of the turbulent flow field under breaking waves in the surf zone. Journal of Fluid Mechanics. 1989;**204**: 359-387

[4] Lakehaland D, Liovic P. Turbulence structure and interaction with steep breaking waves. Journal of Fluid Mechanics. 2011;**674**:522-577

[5] Wei Z, Dalrymple RA, Xu M, Garnier R, Derakhti M. Short-crested waves in the surf zone. Journal of Geophysical Research: Oceans. 2017; **122**. DOI: 10.1002/2016JC012485

[6] Peregrine DH. Breaking waves on beaches. Annual Review of Fluid Mechanics. 1983;**15**:149-178

[7] Basco RD. A qualitative description of wave breaking. J. Waterw. Port Coast. Ocean Eng. ASCE. 1985;**111**:171-188

[8] Battjes JA. Surf-zone dynamics. Annual Review of Fluid Mechanics. 1988;**20**:257-293

[9] Stive, M.J.F. Velocity and pressure field of spilling breakers. In Proceedings of the 17th International Coastal Engineering Conference, Sydney, Australia, 23–28 March 1980; pp. 547–566.

[10] Hattori M, Aono T. Experimental study on turbulence structures under breaking waves. Coastal Engineering. Japan. 1985;**28**:97-116

[11] Ting FCK, King JT. Dynamics of surf-zone turbulence in a strong plunging breaker. Coastal Engineering. 1995;**24**:177-204

[12] Longo S. Turbulence under spilling breakers using discrete wavelets. Experiments in Fluids. 2003;**43**:181-191

[13] Stansby, P.k.; Feng, T. Kinematics and depth-integrated terms in surf zone waves from laboratory measurement. J. Fluid Mech. 2005, 529, 279–310.

[14] De Serio F, Mossa M. Experimental study on the hydrodynamics of regular breaking waves. Coastal Engineering. 2006;**53**:99-113

[15] Sancho F, Mendes PA, Carmo JA, Neves MG, Tomasicchio GR, Archetti R, et al. Wave hydrodynamics over a barred beach. Proceedings of the International Symposium on Ocean Wave Measurement and Analysis. 2001; **2**:1170-1179

[16] De Serio, F.; Mossa, M.A laboratory study of irregulars hoaling waves. Exp. Fluids.2013,54,15–36.

[17] De Padova D, Brocchini M, Buriani F, De Serio F, Mossa M, Sibilla S. Experimental and numerical investigation of pre-breaking and breaking vorticity within a plunging breaker. Water. 2018;**10**(4):387. DOI: 10.3390/w10040387

[18] Wu YT, Hsiao SC, Huang ZC, Hwang KS. Propagation of solitary waves over a bottom-mounted barrier. Coastal Engineering. 2012;**62**:31-47

[19] Chung TJ. Computational Fluid Dynamics. Cambridge: Cambridge University Press; 2002

[20] Anderson JD. Computational Fluid Dynamics: The Basics with Applications. New York: McGraw Hill; 2002

[21] Zienkiewicz OC, Taylor RL. The Finite Element Method. Stonham: Butterworth-Heinemann; 2000

[22] Udaykumar HS, Mittal R, Rampunggoon P, Khanna A. A sharp interface Cartesian grid method for simulating flows with complex moving boundaries. Journal of Computational Physics. 2001;**174**:345-380

[23] Bruno D, De Serio F, Mossa M. The FUNWAVE model application and its validation using laboratory data. Coastal Engineering. July 2009;**56**(7):773-787

[24] Fleissner F, Gaugele T, Eberhard P. Applications of the discrete element method in mechanical engineering. Multibody System Dynamics. 2007;**18**: 81. DOI: https://doi.org/10.1007/s11044-007-9066-2

[25] Liu MB, Liu GR. Smoothed particle hydrodynamics (SPH): An overview and recent developments. Arch Computat Methods Eng. 2010;**17**:25-76. DOI: https://doi.org/10.1007/s11831-010-9040-7

[26] Rabczuk T, Gracie R, Song J-H, Belytschko T. Immersed particle method for fluid-structure interaction. International Journal for Numerical Methods in Engineering. 2010;**22**:48

[27] Rabczuk T, Belytschko T. A three-dimensional large deformation meshfree method for arbitrary evolving cracks. Computer Methods in Applied Mechanics and Engineering. 2007;**196**: 2777-2799

[28] Belytschko T, Krongauz Y, Organ D, Fleming M, Krysl P. Meshless methods: An overview and recent developments. Computer Methods in Applied Mechanics and Engineering. 1996;**139** (1–4):3-47

[29] Liu GR. Meshfree Methods: Moving beyond the Finite Element Method. Boca Raton: CRC Press; 2002

[30] Nguyen VP, Rabczuk T, Bordas S, Duflot M. Meshless methods: A review and computer implementation aspects. Mathematics and Computers in Simulation. 2008;**79**(3):763-813

[31] Dalrymple RA, Rogers BD. Numerical modelling of waves with the SPH method. Coastal Engineering. 2006;**53**(2–3):141-147

[32] De Padova D, Dalrymple RA, Mossa M. Analysis of the artificial viscosity in the smoothed particle hydrodynamics modelling of regular waves. Journal of Hydraulic Research. 2014;**52**(6):836-848

[33] Ni X, Feng WB, Wu D. Numerical simulations of wave interactions with vertical wave barriers using the SPH method. International Journal for Numerical Methods in Fluids. 2014;**76**: 223-245

[34] Lind SJ, Stansby PK, Rogers BD, Lloyd PM. Numerical predictions of water–air wave slam using incompressible– Compressible smoothed particle hydrodynamics. Applied Ocean research. 2015;**49**:57-71

[35] Shadloo MS, Weiss R, Yildiz M, Dalrymple RA. Numerical simulation of long wave run up for breaking and nonbreaking waves. International Journal of Offshore and Polar Engineering. 2015;**25**(1):1

[36] Makris C. V., Memos C.D., Krestenitis Y.N. (2016) Numerical modeling of surf zone dynamics under weakly plunging breakers with SPH method ocean Modelling 98, 12-35

[37] De Padova D, Ben Meftah M, De Serio F, Mossa M, Sibilla S. Characteristics of breaking vorticity in spilling and plunging waves. Environmental Fluid Mechanics. 2020; **20**(2):233-260

[38] De Padova D, Mossa M, Sibilla S. SPH numerical investigation of the velocity field and vorticity generation within a hydrofoil induced spilling breaker. Environmental Fluid Mechanics. 2016;**16**(1):267-287

[39] Manenti S, Sibilla S, Gallati M, Agate G, Guandalini R. SPH simulation of sediment flushing induced by a rapid water flow. Journal of Hydraulic Engineering. 2012;**138**(3):272-284

[40] Manenti S, Pierobon E, Gallati G, et al. Vajont disaster: Smoothed particle hydrodynamics modeling of the post-event 2D experiments. Journal of Hydraulic Engineering. 2016;**142**(4): 1-11

[41] Mokos A, Rogers BD, Stansby PK. Multiphase SPH modelling of violent hydrodynamics on GPUs. Computer Physics Communications. 2015;**196**: 304-316

[42] Ulrich C, Leonardi M, Rung T. Multi-physics SPH simulation of complex marine-engineering hydrodynamic problems. Ocean Engineering. 2013;**64**:109-121

[43] Espa P, Sibilla S, Gallati M. SPH simulations of a vertical 2-D liquid jet introduced from the bottom of a free-surface rectangular tank. Adv. Appl. Fluid Mech. 2008;**3**:105-140

[44] De Padova D, Mossa M, Sibilla S. Characteristics of non buoyant jets in a wave environment investigated numerically by SPH. Environmental Fluid Mechanics. 2020;**20**(1):189-202

[45] De Padova D, Mossa M, Sibilla S. Numerical investigation of the behaviour of jets in awave environment. Journal of Hydraulic Research. 2019. DOI: https://doi.org/10.1080/00221686.2019.1647886

[46] Barile S, De Padova D, Mossa M, Sibilla S. Theoretical analysis and numerical simulations of turbulent jets in awave environment. Physics of Fluids. 2020;**32**(3). DOI: https://doi.org/10.1063/1.5141039

[47] Zhang A, Cao X, Ming F, Zhang ZF. Investigation on a damaged ship model sinking in to water based on three dimensional SPH method. Applied Ocean research. 2013;**42**:24-31

[48] Marrone S, Bouscasse B, Colagrossi A, Antuono M. Study of ship wave breaking patterns using 3D parallel SPH simulations. Computers & Fluids. 2012;**69**:54-66

[49] De Padova D, Mossa M. Modelling fluid–structure interactions: A survey of methods and experimental verification. Proceedings of the Institution of Civil Engineers – Engineering and Computational Mechanics. DOI: https://doi.org/10.1680/jencm.19.00014

[50] De Padova D, Mossa M, Sibilla S, Torti E. 3D SPH modelling of hydraulic jump in a very large channel. Journal of Hydraulic Research. 2013;**51**:158-173

[51] De Padova D, Mossa M, Sibilla S (2017) SPH modelling of hydraulic jump oscillations at an abrupt drop. Water 9 (10):790. https://doi.org/10.3390/w9100790

[52] De Padova D, Mossa M, Sibilla S (2018) SPH numerical investigation of characteristics of hydraulic jumps. Environ Fluid Mech. https://doi.org/10.1007/s10652-017-9566-4

[53] De Padova D, Mossa M, Sibilla S. SPH numerical investigation of the characteristics of an oscillating hydraulic jump at an abrupt drop. Journal of Hydrodynamics. 2018;**30**(1): 106-113

[54] Violeau D, Buvat C, Abed-Meraim K, et al. Numerical modelling of boom and oil spill with SPH. Coastal Engineering. 2007;**54**:895-913

[55] Yang X F, Liu M B. Numerical modeling of oil spill containment by boom using SPH. Sci China-Phys Mech Astron, 2013, 56: 315–321, doi: 10.1007/s11433-012-4980-6

[56] Pope SB. Turbulent Flows. Cambridge: Cambridge University Press; 2000

[57] Quinlan NJ, Basa M, Lastiwka M. Truncation error in mesh free particle methods. International Journal for Numerical Methods in Engineering. 2006;**66**(13):2064-2085

[58] Randles P, Libersky L. Smoothed particle hydrodynamics: Some recent improvements and applications. Computer Methods in Applied Mechanics and Engineering. 1996; **139**(1):375-408

[59] Bonet J, Lok T-SL. Variational and momentum preservation aspects of smoothed particle hydrodynamic formulations. Computer Methods in Applied Mechanics and Engineering. 1999;**180**:97-115

[60] Dehnen W, Aly H. Improving convergence in smoothed particle hydrodynamics simulations without pairing instability. Monthly Notices of the Royal Astronomical Society. 2012; **425**(2):1068-1082

[61] Wendland H. Piecewise polynomial, positive definite and compactly supported radial functions of minimal degree. Advances in Computational Mathematics. 1995;**4**:389-396

[62] Liu GR, Liu MB. Smoothed particle hydrodynamics (SPH): An overview and recent developments. Archives of Computational Methods in Engineering. 2010;**17**:25

[63] Morris JP. A study of the stability properties of smooth particle hydrodynamics. Publications of the Astronomical Society of Australia. 1996; **13**(1):97-102

[64] Monaghan JJ, Lattanzio JC. A refined particle method for astrophysical problems. Astronomy and Astrophysics. 1985;**149**(1):135-143

[65] Espa P, Sibilla S, Gallati M. SPH simulations of avertical 2-D liquid jet introduced from the bottom of a free-surface rectangular tank. Adv. Appl. Fluid Mech. 2008;**3**:105-140

[66] Manenti S, Sibilla S, Gallati M, Agate G, Guandalini R. SPH simulation of sediment flushing induced by a rapid water flow. Journal of Hydraulic Engineering. 2012;**138**(3):272-284

[67] Monaghan JJ. Smoothed particle hydrodynamics. Reports on Progress in Physics. 2005;**68**:1703-1759

[68] Gomez-Gesteira M, Rogers BD, Darlymple RA, Crespo AJC. State-of-the-art of classical SPH for freesurface flows. Journal of Hydraulic Research. 2010;**48**:6-27

[69] Sibilla S. An algorithm to improve consistency in smoothed particle hydrodynamics. Computers and Fluids. 2015;**118**:148-158

[70] Monaghan JJ. Smoothed particle hydrodynamics. Annual Review of Astronomy and Astrophysics. 1992;**30**: 543-574

[71] Sibilla S. SPH simulation of local scour processes. ERCOFTAC Bulletin. 2008;**76**:41-44

[72] Colagrossi A, Landrini M. Numerical simulation of interfacial flows by smoothed particle hydrodynamics. Journal of Computational Physics. 2003;**191**(2): 448-475

[73] Antuono M, Colagrossi A, Marrone S, Molteni D. Free-surface

flows solved by means of SPH schemes with numerical diffusive terms. Computer Physics Communications. 2010;**181**:532-549

[74] Colagrossi A, Landrini M. Numerical simulation of interfacial flows by smoothed particle hydrodynamics. Journal of Computational Physics. 2003;**191**(2): 448-475

[75] Wang D, Liu PL-F. An ISPH with k–ε closure for simulating turbulence under solitary waves. Coastal Engineering. 2020;**157**:103657. DOI: https://doi.org/10.1016/j.coastaleng.2020.103657

[76] Dean R.G. and Dalrymple R.A. Water Wave Mechanics for Engineers and Scientists. World Scientific (1991)

[77] Peregrine, D.H., Svendsen, I.A., 1978. Spilling breakers, bores and hydraulic jumps. In: Proceedings of the 16th International Conference on Coastal Engineering. ICCE. Hamburg. ASCE, pp. 540–550.

[78] Zhao Q, Armfield S, Tanimoto K. Numerical simulation of breaking waves by a multi-scale turbulence model. Coastal Engineering. 2004;**51**:53-80

[79] Misra SK, Kirby JT, Brocchini M, Veron F, Thomas M, Kambhamettu C. The mean and turbulent flow structure of a weak hydraulic jump. Physics of Fluids. 2008;**20**:035106

[80] Dabiri D, Gharib M. Experimental investigation of the vorticity generation within a spilling water wave. Journal of Fluid Mechanics. 1997;**330**:113-139